Ceramic Materials for
Energy Applications III

Ceramic Materials for Energy Applications III

*A Collection of Papers Presented at the
37th International Conference on
Advanced Ceramics and Composites
January 27–February 1, 2013
Daytona Beach, Florida*

Edited by
Hua-Tay Lin
Yutai Katoh
Alberto Vomiero

Volume Editors
Soshu Kirihara
Sujanto Widjaja

The
American
Ceramic
Society

WILEY

Cover Design: Wiley

Published by John Wiley & Sons, Inc., Hoboken, New Jersey.
Published simultaneously in Canada.

For general information on our other products and services or for technical support, please contact our
Customer Care Department within the United States at (800) 762-2974, outside the United States at
(317) 572-3993 or fax (317) 572-4002.

Wiley also publishes its books in a variety of electronic formats. Some content that appears in print may
not be available in electronic formats. For more information about Wiley products, visit our web site at
www.wiley.com.

Library of Congress Cataloging-in-Publication Data is available.

ISBN: 978-1-118-80758-3
ISSN: 0196-6219

Printed in the United States of America.

10 9 8 7 6 5 4 3 2 1

Contents

ADVANCED CERAMIC MATERIALS AND PROCESSING FOR PHOTONICS AND ENERGY

ADVANCED CERAMICS AND COMPOSITES FOR SUSTAINABLE NUCLEAR ENERGY AND FUSION ENERGY

Preface

This proceedings issue contains contributions from three energy related symposia and the Engineering Ceramics Summit of the America that were part of the 37th International Conference on Advanced Ceramics and Composites (ICACC), in Daytona Beach, Florida, January 27-February 1, 2013. The symposia include Advanced Materials and Technologies for Energy Generation and Rechargeable Energy Storage; Advanced Ceramics and Composites for Sustainable Nuclear Energy and Fusion Energy; and Advanced Ceramic Materials and Processing for Photonics and Energy. These symposia and the Summit were sponsored by the ACerS Engineering Ceramics Division. The symposium on Advanced Ceramics and Composites for Sustainable Nuclear Energy and Fusion Energy was cosponsored by the ACerS Nuclear and Environmental Technology Division.

The editors wish to thank the authors and presenters for their contributions, the symposium organizers for their time and labor, and all the manuscript reviewers for their valuable comments and suggestions. Acknowledgment is also due for financial support from the Engineering Ceramics Division, the Nuclear and Environmental Technology Division, and The American Ceramic Society. The editors wish to thank Greg Geiger at ACerS for all his effort in assembling and publishing the proceedings.

HUA-TAY LIN, *Oak Ridge National Laboratory, USA*
YUTAI KATOH, *Oak Ridge National Laboratory, USA*
ALBERTO VOMIERO, *CNR–University of Brescia, Italy*

Introduction

This issue of the Ceramic Engineering and Science Proceedings (CESP) is one of nine issues that has been published based on manuscripts submitted and approved for the proceedings of the 37th International Conference on Advanced Ceramics and Composites (ICACC), held January 27–February 1, 2013 in Daytona Beach, Florida. ICACC is the most prominent international meeting in the area of advanced structural, functional, and nanoscopic ceramics, composites, and other emerging ceramic materials and technologies. This prestigious conference has been organized by The American Ceramic Society's (ACerS) Engineering Ceramics Division (ECD) since 1977.

The 37th ICACC hosted more than 1,000 attendees from 40 countries and approximately 800 presentations. The topics ranged from ceramic nanomaterials to structural reliability of ceramic components which demonstrated the linkage between materials science developments at the atomic level and macro level structural applications. Papers addressed material, model, and component development and investigated the interrelations between the processing, properties, and microstructure of ceramic materials.

The conference was organized into the following 19 symposia and sessions:

Symposium 1	Mechanical Behavior and Performance of Ceramics and Composites
Symposium 2	Advanced Ceramic Coatings for Structural, Environmental, and Functional Applications
Symposium 3	10th International Symposium on Solid Oxide Fuel Cells (SOFC): Materials, Science, and Technology
Symposium 4	Armor Ceramics
Symposium 5	Next Generation Bioceramics
Symposium 6	International Symposium on Ceramics for Electric Energy Generation, Storage, and Distribution
Symposium 7	7th International Symposium on Nanostructured Materials and Nanocomposites: Development and Applications

Symposium 8	7th International Symposium on Advanced Processing & Manufacturing Technologies for Structural & Multifunctional Materials and Systems (APMT)
Symposium 9	Porous Ceramics: Novel Developments and Applications
Symposium 10	Virtual Materials (Computational) Design and Ceramic Genome
Symposium 11	Next Generation Technologies for Innovative Surface Coatings
Symposium 12	Materials for Extreme Environments: Ultrahigh Temperature Ceramics (UHTCs) and Nanolaminated Ternary Carbides and Nitrides (MAX Phases)
Symposium 13	Advanced Ceramics and Composites for Sustainable Nuclear Energy and Fusion Energy
Focused Session 1	Geopolymers and Chemically Bonded Ceramics
Focused Session 2	Thermal Management Materials and Technologies
Focused Session 3	Nanomaterials for Sensing Applications
Focused Session 4	Advanced Ceramic Materials and Processing for Photonics and Energy
Special Session	Engineering Ceramics Summit of the Americas
Special Session	2nd Global Young Investigators Forum

The proceedings papers from this conference are published in the below nine issues of the 2013 CESP; Volume 34, Issues 2–10:

- Mechanical Properties and Performance of Engineering Ceramics and Composites VIII, CESP Volume 34, Issue 2 (includes papers from Symposium 1)
- Advanced Ceramic Coatings and Materials for Extreme Environments III, Volume 34, Issue 3 (includes papers from Symposia 2 and 11)
- Advances in Solid Oxide Fuel Cells IX, CESP Volume 34, Issue 4 (includes papers from Symposium 3)
- Advances in Ceramic Armor IX, CESP Volume 34, Issue 5 (includes papers from Symposium 4)
- Advances in Bioceramics and Porous Ceramics VI, CESP Volume 34, Issue 6 (includes papers from Symposia 5 and 9)
- Nanostructured Materials and Nanotechnology VII, CESP Volume 34, Issue 7 (includes papers from Symposium 7 and FS3)
- Advanced Processing and Manufacturing Technologies for Structural and Multi functional Materials VII, CESP Volume 34, Issue 8 (includes papers from Symposium 8)
- Ceramic Materials for Energy Applications III, CESP Volume 34, Issue 9 (includes papers from Symposia 6, 13, and FS4)
- Developments in Strategic Materials and Computational Design IV, CESP Volume 34, Issue 10 (includes papers from Symposium 10 and 12 and from Focused Sessions 1 and 2)

The organization of the Daytona Beach meeting and the publication of these proceedings were possible thanks to the professional staff of ACerS and the tireless dedication of many ECD members. We would especially like to express our sincere thanks to the symposia organizers, session chairs, presenters and conference attendees, for their efforts and enthusiastic participation in the vibrant and cutting-edge conference.

ACerS and the ECD invite you to attend the 38th International Conference on Advanced Ceramics and Composites (http://www.ceramics.org/daytona2014) January 26-31, 2014 in Daytona Beach, Florida.

To purchase additional CESP issues as well as other ceramic publications, visit the ACerS-Wiley Publications home page at www.wiley.com/go/ceramics.

SOSHU KIRIHARA, *Osaka University, Japan*
SUJANTO WIDJAJA, *Corning Incorporated, USA*

Volume Editors
August 2013

Energy Summit
of the Americas

NEW MATERIALS FOR ENERGY AND BIOMEDICAL APPLICATIONS

Alejandra Hortencia Miranda González [2], Claudio Machado Junior [2], Bruna Andressa Bregadiolli [1], Natália Coelho de Farias [2], Paulo Henrique Perlatti D'Alpino [2], Carlos Frederico de Oliveira Graeff [1]

[1] Departamento de Física, Faculdade de Ciências, UNESP – Universidade Estadual Paulista, Av. Eng. Luiz Edmundo Carrijo Coube, 14-01, CEP 17033-360, Bauru, SP, Brazil
[2] Biomaterials Research Group, School of Dentistry, Universidade Anhanguera-UNIBAN, Rua Maria Cândida, 1813, CEP 02071-013, São Paulo, SP, Brazil

ABSTRACT

We have conducted research works on the synthesis and characterization of ceramic materials for energy technologies and the characterization of bioceramics and composites for dental applications. $La_{0.50}Li_{0.50}TiO_3$ nanoparticles were synthesized by the polymeric precursor method (PPM) for application as cathode in secondary lithium batteries. Electrochemical measurements of reduction/oxidation processes gave evidence of two kinetic processes. Also, nanocrystalline TiO_2 films for application in hybrid solar cells were prepared. The particles were prepared from a "sol" solution by microwave assisted hydrothermal synthesis. The films were deposited by painting on ITO substrate and sintered at 450°C. XRD characterization indicated the crystallization of anatase phase. In addition, the preparation of ZrO_2 bioceramic by PPM was investigated. Structural characterization by XRD showed that the powders are polycrystalline and free of secondary phases. Morphological results revealed a microstructure of spherical grains with homogeneous sizes of 70 nm. Finally, the ability to accurately predict changes in dental composite properties is of critical importance for the industry, researchers, and clinicians. An accelerated aging process has been used to characterize the mechanical, structural and rheological parameters of composites. Results have shown that this process influences most of the parameters when predicting 9 months of aging.

INTRODUCTION

The main research activities of the *Departamento de Física* (DF, Dep. of Physics) of the *Faculdade de Ciências* (FC, Science Faculty) of *Universidade Estadual Paulista* (UNESP) campus of Bauru is on Materials Science and Technology. Of its 24 lectures 9 belongs to the Graduate College in Materials Science and Technology - POSMAT. POSMAT is a graduate college in network comprising 7 different units in 6 different campus of UNESP. The activities of POSMAT are focused on interdisciplinary and multidisciplinary research, with the mission of development of basic and applied research, promoting a systematic transfer of this knowledge, for technological application and educational purposes. The Laboratory of New Materials and Devices (LNMD), one of the research groups of DF-FC, does research on ceramic materials as well as in organic semiconductors. LNMD provides the opportunity for scientists of different areas such as theoreticians and experimentalists, physicists, chemists and dentists to discuss and work together, thereby creating new perspectives in technological research.

Research at the LNMD regarding ceramic materials is focused in the synthesis and characterization of new materials for energy technologies and bioceramic application, as well as the characterization of composites for dental applications.

Considering energy technologies, it is unquestionable that global warming, the exhaustion of fossil fuels, and the need to prevent air pollution demand the urgent development of environmentally friendly energy sources. Among several alternatives, secondary lithium batteries are promising candidates.[1] Of special interest are lithium lanthanum titanates (LLTO), with the general formula $La_{2/3-x}Li_{3x}TiO_3$, because these compounds are among the best conductors containing Li^+ ion known to date, and therefore are

3

promising candidates for use as solid electrolytes in lithium secondary batteries, especially as cathodes.[2] Recently, there has been considerable growing interest in the synthesis of materials for active cathodes from chemical routes based on aqueous solutions.[3] Of special interest for the synthesis of $La_{0.50}Li_{0.50}TiO_3$ for use in cathodes is the polymeric precursor method, also known as the Pechini Method.[4] In this context, the aim of the present study was to prepare $La_{0.50}Li_{0.50}TiO_3$ nanoparticles by means of the PPM. As characterization techniques, X-ray powder diffraction (XRD), scanning electron microscopy-field emission gun (FEG-SEM), and chronopotentiometry were used.

As sources of energy, third generation solar cells fabricated using hybrid and organic materials are a promising alternative to the expensive commercial silicon solar cells. Known as "Grätzel" solar cells[5], the dye sensitized solar cells (DSSC) have received special attention due to their easy of processing, low cost and good performance.[6] This device works when a photon absorbed by a dye molecule gives rise to electron injection into the conduction band of the semiconductor, in most cases TiO_2, to generate the current. To complete the circuit, the dye must be regenerated by an electron transfer from the electrolyte, which is then reduced at the counter electrode. For a better performance, the TiO_2 has to have large superficial area, porosity and adherence to the conductor substrate.[5,7] Of special interest for the synthesis of TiO_2 in DSSC is the microwave assisted hydrothermal technique.[8] In this context, the aim of the present study was to prepare TiO_2 nanoparticles by means of this method. X-ray powder diffraction (XRD), scanning electron microscopy-field emission gun (FEG-SEM), and transmission electron microscopy (TEM) were used to characterize the obtained material.

The development of zirconia (ZrO_2) nanoparticles has attracted much attention due to their multifunctional characteristics.[9] Considering the application as a biomaterial, the development of advanced dental material technologies has recently led to the use of zirconia-based ceramics in dentistry. The remarkable mechanical properties of zirconia are mainly due to the tetragonal to monoclinic (t → m) phase transformation. The t → m transformation, which can be induced by external stresses, such as grinding, cooling and impact, results in a 4% increase of volume that causes compressive stresses. These stresses may develop on the surface or in the vicinity of a tip crack.[10] The presence of tetragonal phase is an essential condition for zirconia toughening besides hindering or interrupting crack propagation.[11] With the advent of nanotechnology, several techniques have been employed to obtain ceramic powders with nanometric dimensions from chemical processes. Chemical synthesis allows the manipulation of matter at the molecular level, enabling good chemical homogeneity, and allows for the control of particle size and shape.[12] In this context, the aim of the present study was to prepare ZrO_2 nanopowders directly through the polymeric precursor method and investigate the influence of heat treatment on the structure of powders.

Another research field developed at LNMD comprises the chemical, structural, and mechanical characterization of dental composites subjected to accelerated aging processes. Dental composites are subjected to a variety of chemical challenges in the oral cavity. Composites are exposed to water and chemicals provided by acidic food and beverages that modify the pH of the saliva. These events create a variety of chemical and physical processes that produce deleterious effects on the structure and function of dental resin composites. In an aqueous environment, composites absorb water that not only elutes unreacted monomers, but also filler particles. Clinically, the release of composite components influences the initial dimensional changes of the composite as well as the longevity of resin-based composite restorations. The sorption of water may lead to the dissolution of the polymer matrix and/or entrapment of water molecules in polymer voids, by means of chemical bonds scission in the resin or softening through the plasticizing action of water.[13] The great majority of dental composites are methacrylate-based materials that contain pendant hydroxyl groups in their molecular backbone. Because of these polar groups, polymers made with these monomers tend to be somewhat hydrophilic and susceptible to water sorption..

Recently new composite formulations were launched and most of them are methacrylate-based. The introduction of new or modified composites to the market requires the characterization of these materials to assure that their use can be made with safety and efficacy. Composites are made not only of monomers, but also by filler particles, plasticizers, colorants, photo-initiators, and stabilizers. The functional properties of a dental composite depend on the properties of its components. The different compositions associated with the different clinical uses (handling of the materials, application technique, polymerization technique, finishing and polishing, among others) determines the degradation mechanism of the dental resin composite.[14] In order to obtain experimental data on performance and shelf-life of these composites, accelerated-aging tests are used. Most of these tests are made using higher temperatures based on the assumption that reaction rates are increased exponentialy with T.[15] Another relevant issue is that the properties of any new product may vary as a function of the manufacturing process, transport, storage conditions, among others, until its use in the dental clinic. In other words, how degraded are these new products, when they reach the dentist?

EXPERIMENTAL

Energy

The Polymeric Precursor Method (PPM) was employed for preparation of the $La_{0.50}Li_{0.50}TiO_3$ (LLTO) powder. This method is based on metallic citrate polymerization with the use of ethylene glycol. Polymerization, promoted by heating of the mixture, results in a homogeneous resin in which the metal ions are uniformly distributed throughout the organic matrix. Titanium (IV) isopropoxide, $Ti[OCH(CH_3)_2]_4$ (Alfa Aesar), lithium nitrate, $LiNO_3$ (Vetec), and lanthanum nitrate, $La(NO_3)_3.6H_2O$ (Vetec) were utilized as raw materials. Titanium (IV) isopropoxide was added with stirring to ethylene glycol (60°C), and then citric acid was added to this solution. The metal/citric acid/ethylene glycol molar ratio was set at 1:4:16. Stoichiometric quantities of lithium and lanthanum nitrates were added to the titanium solution. The mixture was kept stirring and heated up to 130°C in order to accelerate esterification reactions between citric acid and ethylene glycol. The prolonged heating at this temperature produced a viscous transparent resin. Charring the resin at 350°C for 3 h in a box furnace resulted in a black solid mass, the precursor powder. The powder was finally heated from 600°C to 700°C for 3 h, in static air, to reach the crystallization stage.

The thermal evolution of the precursor powder heated from 350°C to 700°C was followed by means of X-ray powder diffraction (XRD) carried out using a Rigaku RINT2000 diffractometer (Cu k_α radiation), in the 2θ range between 10° and 70° with an interpolated step of 0.02°. The LLTO microstructure was investigated using a field emission gun-scanning electron microscope (FEG-SEM), FEG-VP Zeiss Supra 35. The electrochemical characterization was accomplished by charge–discharge tests. Intercalation and deintercalation of lithium in the LLTO were performed using a Swagelok cell with lithium as anode and LLTO as active material. The working electrode was a $La_{0.50}Li_{0.50}TiO_3$ pellet with a diameter of 12 mm and a thickness of 1 mm, obtained at a pressure of 1 ton. The mass of electroactive material was 100 mg. The electrolyte employed in this work was 1 mol L^{-1} $LiClO_4$ in a 50:50 (w/w) mixture of ethylene carbonate (EC) and dimethyl carbonate (DMC). The cell was mounted in an argon glove box (Mbraun, Unilab, Germany), to prevent contamination with water and oxygen. A Celgard® porous propylene membrane soaked with the electrolyte was used to separate the working and counter electrodes. Charge and discharge tests were conducted with a voltage limit between 4.6 and 0.9 V at a constant current of 100 μA. For these measurements an Autolab PGSTAT30 Potenciostat, controlled by the GPES software was used. All electrochemical experiments were done at room temperature (RT).

The Microwave Assisted Hydrothermal Technique was used for the synthesis of the TiO_2 powder. This method is based on microwave heating of the precursor solution in a close vessel that as a consequence reaches predefined pressures and temperatures. Titanium (IV) isopropoxide, Ti[OCH(CH$_3$)$_2$]$_4$ (Alfa Aesar), nitric acid (Dinâmica) and distilled water were utilized as raw materials. Titanium (IV) isopropoxide was added with the acid to water and stirred at 60°C. The mixture was kept stirring for 12 h. The prolonged heating at this temperature produced a homogeneous resin in which the metal ions are uniformly distributed throughout the solution. This "sol" solution was put into the autoclave and treated in different temperatures from 110 to 150°C from 15 to 60 min. The thermal influence of the treatment of the precursor solution from 110°C to 150°C was followed by means of X-ray powder diffraction (XRD) carried out using the same equipment and conditions previously described. The TiO$_2$ microstructure was investigated using a FEG-VP Zeiss Supra 35 field emission gun scanning electron microscope (FEG-SEM), and a Philips CM120 transmission electron microscope (TEM). The superficial area was evaluated by adsorption of N$_2$ at 77 K (S$_{BET}$) using a Micrometrics ASAP 2010. After the characterization, the concentration of the solutions was adjusted to 200 mg / mL with addition of 40 wt% of polyethyleneglycol (20,000 MW, Sigma Aldrich) and stirred for 8 h at RT. The paste obtained was deposited by the painting method on ITO substrate and sintered at 450°C.

Biomedical

To synthesize the zirconium precursor source by PPM, a solution of 80% in mass of zirconium butoxide in 1-butanol, Zr(OC$_4$H$_9$)$_4$ (Aldrich) was used. The preparation involved the following steps: first, 19.45 g of the zirconium butoxide solution was added to a beaker containing 100 mL of deionized water. This solution was kept under constant stirring at 80°C for 2 h to obtain a Zr(OH)$_4$ suspension. Citric acid anhydride (Synth) was then added to the metallic ion complex, establishing a citric acid:Zr^{4+} metallic ion molar ratio of 4:1. The solution was kept under stirring at a temperature of 80°C for 48 h to trigger the complexation process, after which ethyl glycol was added and the solution kept at 80°C to produce the esterification and polymerization reaction. The metal/ethylene glycol molar ratio was set at 1:16. The prolonged heating at 130°C produced a viscous transparent resin. The effect of high-temperature treatments from 350°C to 800°C for 3 h on the structure of powders was monitored using powder X-ray diffraction (XRD). In this case, the 2θ range was between 10° and 80° with an interpolated step of 0.02°. Morphology and grain size of the powder heat treated at 800°C was investigated by means of field emission gun scanning electron microscopy (FEG-SEM).

The composites for dental application, syringes of 3M ESPE Filtek Silorane were submitted to an accelerated aging process at 37°C for 12 weeks, simulating an aging protocol of 9 months (Aged). Rectangular specimens were prepared (8.0 x 2.0 x 2.0 mm). After photoactivation for either 20 or 40 s (Bluephase, 1200 mW/cm^2), the specimens were stored in distilled water at 37°C for 24 h. The specimens were subsequently blot-dried, measured and subjected to mini-flexural testing (0.75 mm/min). The flexural strength and the flexural modulus were calculated. The pre-curing rheological parameters were also obtained by using 0.5g of the composite diluted in ethanol p.a. in a rheometer (Brookfield DV-III Ultra Rheometer). The same measurements were made in pure ethanol to obtain a baseline curve. Values of η (cP) and yield stress (D/cm^2) were calculated. The results were compared to that obtained with a new syringe of a non-aged silorane composite (New) and also with another syringe of expired silorane composite (Expired), both from the same brand. Data were analyzed using ANOVA/Tukey's test (α = 5%).

RESULTS AND DISCUSSION

Energy

Figure 1 shows XRD patterns of the lithium lanthanum titanate precursor powder annealed from 350°C to 700°C for 3 h. The formation of the crystalline phase with the increase in calcination temperatures can be observed. A correlation between the process of crystallization and organic fraction elimination is therefore evident. A highly crystalline perovskite phase has been identified for the samples heat-treated at 700°C.

All peaks and associated planes represented in Figure 1 are in agreement with the crystallographic data of the $Li_{0.34}La_{0.51}TiO_{2.94}$ phase described by Inaguma et al.[16] However, a careful analysis of the XRD patterns shows that the LLTO formation starts at 700°C but with impurities. The peak at $2\theta = 28.8°$ could be related to the formation of $La_2Ti_2O_7$. This was also found by Bohnke et al.[17] during the preparation of $Li_{0.33}La_{0.56}TiO_3$ by the sol-gel method. The second phase marked with asterisks ($2\theta = 18.5°$ and $43.5°$) can be attributed to the formation of $Li_2Ti_2O_5$, according to Belous et al.[18] The co-existence of secondary phases with LLTO implies that the stoichiometry of LLTO has been modified.

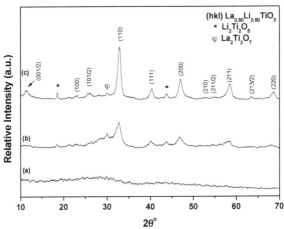

Figure 1. XRD patterns of $La_{0.50}Li_{0.50}TiO_3$ samples thermally treated at: (a) 350°C/3h; (b) 600°C/3h; (c) 700°C/3h.

Figure 2 presents the FEG-SEM micrographs of the $La_{0.50}Li_{0.50}TiO_3$ powder heat-treated at 700°C for 3 h, at different magnifications. It can be seen that the powder consists of spherical particles with homogeneous sizes in the order of 20–30 nm, Figure 2a. However, these particles agglomerate, as shown in Figure 2b.

Intercalation and deintercalation of lithium in $La_{0.50}Li_{0.50}TiO_3$ were performed by means of the chronopotentiometry technique. Figure 3 displays the preliminary results of charge (deintercalation) and discharge (intercalation) curves obtained by application of a current of 100 µA to LLTO. It is known that the Li intercalation rate per mole of LLTO is low. Moreover, theoretical calculations predict that it is necessary to keep the cell discharging for at least 25 h for the studied composition to complete the process. This time period corresponds to the intercalation of 0.16 Li per mole of LLTO (i.e., 25.4 mAhg^{-1}). However, only 2 h were used for each process in our case. In this condition, one can be sure that the total intercalation was not achieved. During the charge and discharge processes, the LLTO gives rise to extremely flat curves at about 1.98 V, which means that there is a two-phase reaction based on the Ti^{+4}/Ti^{+3} redox couple.[19, 20] After twenty charge–discharge cycles, two different steps in the charge process can be identified, probably related to two different kinetic processes.

(a)   (b)

Figure 2. FEG-SEM micrographs of the $La_{0.50}Li_{0.50}TiO_3$ powder obtained by the Pechini method after heating at 700°C for 3 h. Magnification: (a) 250.000x and (b) 50.000x.

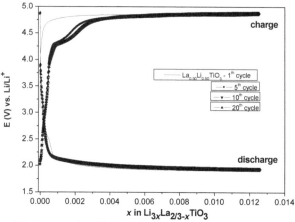

Figure 3. Charge/discharge cycles of $Li/LiClO_4$ (EC:DMC)/$La_{0.50}Li_{0.50}TiO_3$ applying a current of 100 μA.

Figure 4 shows XRD patterns of titanium dioxide powder synthesized at different conditions. It can be seen that the patterns are all similar. The diffraction patterns are compatible with the formation of the anatase phase.[21] However, there is an unidentified peak at $2\theta = 31.8°$ represented by (*) that can be attributed to (211) plane of the brookite phase.[22] The formation of a metastable brookite phase can be attributed to the high pressures inside the reactor, which can reach up to 6 atm. Thus it is possible to conclude from XRD that the reaction temperature is not accompanied by significant changes in the structure of the material.

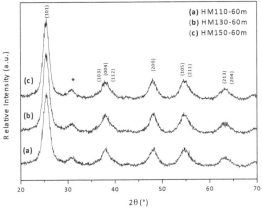

Figure 4. XRD patterns of TiO₂ powders synthesized by microwave assisted hydrothermal technique at different temperatures (a) 110°C (b) 130°C and (c) 150°C. The synthesis time was 60 min.

Figure 5 presents the FEG-SEM and TEM micrographs of the titanium dioxide powder synthesized by microwave assisted hydrothermal technique at 150°C for 60 min. It can be seen in Figure 5a that the powder consists of an agglomerate of particles with variable sizes. For a better morphological analysis of these particles TEM images were carried out, shown in Figure 5b. As can be seen particles have sizes varying between 5 and 15 nm, and form agglomerates. The image can be interpreted as an agglomerate of many nanoparticles or a polycrystalline nanoparticle.

(a)   **(b)**

Figure 5. Micrographs of TiO₂ powder obtained at 150°C for 60 min (a) SEM and (b) TEM.

The surface area of the samples was evaluated by adsorption of N_2 at 77 K, and found to be 207.6 ± 0.9 m² / g. This value is consistent with previous studies[23] and ideal for application in dye solar cells.

Biomedical
Figure 6 shows normalized XRD patterns of zirconia powder annealed from 350°C to 800°C for 3 h. The XRD patterns of all powders presents the characteristic peaks of the tetragonal crystal phase. However, for the powders treated above 500°C/3 h the monoclinic phase is also

observed. The relative intensity of the diffraction peaks increases with increasing anneling temperature in the range of 350°C to 700°C, not shown. This suggests that the cristallinity is improved with increasing annealing temperature. However, heating the powder at 800°C led to the formation of highly crystalline monoclinic phase, with consequent reduction of the tetragonal phase.

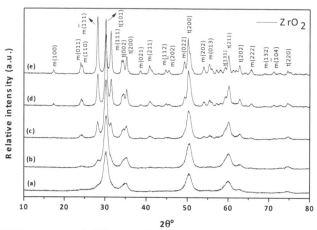

Figure 6. XRD patterns of ZrO_2 samples thermally treated for 3 h at: (a) 350°C; (b) 500°C; (c) 600°C; (d) 700°C; (e) 800°C.

Figure 7 presents the FEG-SEM micrograph of the ZrO_2 powder heat-treated at 800°C for 3 h. Morphological result revealed a microstructure of spherical particles with homogeneous sizes of 30-70 nm.

Figure 7. FEG-SEM micrograph of the ZrO_2 powder synthesized by the polymeric precursor method after heating at 800°C for 3 h.

Tables 1 shows the results of the flexural strength and modulus for silorane composites after different treatments.

	Control		Aged		Expired	
Energy dose	Flexural Strength	Flexural Modulus	Flexural Strength	Flexural Modulus	Flexural Strength	Flexural Modulus
20 s	236.8 (43.1)[a,A]	16.4	294.4 (43.7)[a,A]	13.7	214.6 (46.4)[a,A]	15.7
40 s	207.1 (28.9)[a,A]	14.4	291.4 (59.4)[b,A]	13.8	181.4 (71.8)[a,A]	13.0

Table 1: Flexural Strength and Modulus for silorane composites after different treatments.

The average flexural strength of experimental groups was statistically equivalent, with the exception of Aged/40 s that presented a significantly higher average ($p<0.05$). No significant change was noted when the average flexural moduli were compared ($p>0.05$). The results of η and yield stress are shown in Table 2 and Figure 8. As can be seen the the yield stress is strongly affected by the aging treatment.

	η (cP)	Yield stress (τ°) (D/cm^2)	Confidence of fit-value (%)
Control	1.6	0.07	94.3
Aged	1.5	0.15	94.7
Expired	1.4	0.01	94.5

Table 2: Viscosity and yield stress for silorane composites after different treatments.

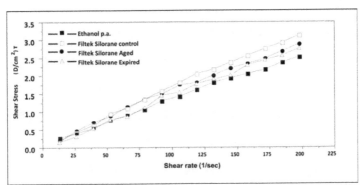

Figure 8. Curves of shear stress x shear rate for silorane composites after different treatments.

SUMMARY AND CONCLUSIONS

The polymeric precursor method was used as an alternative method for the synthesis of La$_{0.50}$Li$_{0.50}$TiO$_3$. Characterization by XRD indicated that a highly crystalline perovskite phase was obtained for the samples heat-treated at 700°C. This temperature is much lower than what is found in conventional methods where temperatures above 1100°C are needed. However, the LLTO crystallization process was accompanied by crystallization of secondary phases, identified as $La_2Ti_2O_7$ and $Li_2Ti_2O_5$. Thus, to obtain pure LLTO using the PPM, new studies are required. The properties of the LLTO powders obtained in this work are comparable to those of similar materials prepared by the sol–gel and other solid-state reaction methods. The charge–discharge curves obtained for the LLTO material produced herein revealed that the lithium ion intercalation and deintercalation processes are reversible in the $La_{0.50}Li_{0.50}TiO_3$ structure. During the cycling process (charge process), two different steps were observed, probably due to two different kinetic processes.

It has been found that the anatase TiO_2 can be prepared by microwave assisted hydrothermal technique; however with partial crystallization of the brookite phase. The results indicate that the synthesis temperature does not have great influence on the final morphology and the crystalline phase of the powders. The particles obtained have a high surface area which ensures that they are ideal for dye solar cells applications.

Zirconia nanopowders were successfully prepared by the polymeric precursor method. According to the XRD results, the ZrO_2 powders are polycrystalline and secondary phases free. A highly crystalline tetragonal phase has been identified for the sample heat-treated in the range of 350-700°C. However, the thermal treatment at 800°C promoted the crystallization of both tetragonal and monoclinic phases. Morphological results revealed a microstructure of spherical particles with sizes ranging between 30-70 nm.

The aging process of silorane composites causes an increase in the flexural strength when the composite is photoactivated for 40 s. The accelerated aging protocol used has no influence in the flexural modulus, but influences the rheology of the material tested. The uncured composites are more affected by the aging treatment.

ACKNOWLEDGMENTS

The authors are grateful to the following Brazilian funding agencies: Fundação de Amparo à Pesquisa do Estado de São Paulo (FAPESP), grants 05/58446-8, 11/11966-8 and 11/02205-3 and Conselho Nacional de Desenvolvimento Científico e Tecnológico (CNPq), grants 479744/2010-6 and 163102/2011-2.

REFERENCES
[1]B. Antoniassi, A. H. M. González, S. L. Fernandes, C. F. O. Graeff, Microstructural and electrochemical study of $La_{0.5}Li_{0.5}TiO_3$, Mater. Chem. Phys., 127, 51-55 (2011).
[2]Y. J. Shan, L. Chen, Y. Inaguma, M. Itoh, T. Nakamura, Oxide cathode with perovskite structure for rechargeable lithium batteries, J. Power Sources, 54, 397-402 (1995).
[3]G. T .K. Fey, R. F. Shiu, T. P. Kumar, C. L. Chen, Preparation and characterization of lithium nickel cobalt oxide powders via a wet chemistry processing, Mater. Sci. Eng. B - Solid, 100, 234-243 (2003).
[4]M. P. Pechini, Method of preparing lead and alkaline earth titanates and niobates and coating method using the same to form a capacitor, US Patent 3330697 (1967).
[5]B. O'Regan, M. Gratzel, A low-cost, high-efficiency solar cell based on dye-sensitized colloidal TiO_2 films, Nature, 353, 737-739 (1991).
[6]S. E. Shaheen, D. S. Ginley, G. E. Jabbour, Organic-based photovoltaics: toward low-cost power generation, MRS Bulletin, 30, 10-15 (2005).
[7]M. Grätzel, Dye-sensitized solar cells, J. Photoch. Photobio. C, 4, 145-153 (2003).
[8]C. H. Huang, Y. T. Yang, R. A. Doong, Microwave-assisted hydrothermal synthesis of mesoporous anatase TiO_2 via sol-gel process for dye-sensitized solar cells, Micropor. Mesopor. Mat., 142, 473-480 (2011).
[9]R. Dwivedi, A. Maurya, A. Verma, R. Prasad, K. S. Bartwal, Microwave assisted sol-gel synthesis of tetragonal zirconia nanoparticles, J. Alloy Compd., 509, 6848-6851 (2011).

[10]M. Guazzato, M. Albakry, S. P. Ringer, M. V. Swain, Strength, fracture toughness and microstructure of a selection of all-ceramic materials. Part II. Zirconia-based dental ceramics, *Dent. Mater.*, **20**, 449-456 (2004).

[11]I. Denry, J. R. Kelly, State of the art of zirconia for dental applications, *Dent. Mater.*, **24**, 299-307 (2008).

[12]M. Bhagwat, V. Ramaswamy, Synthesis of nanocrystalline zirconia by amorphous citrate route: structural and thermal (HTXRD) studies, *Mater. Res. Bull.*, **39**, 1627-1640 (2004).

[13]N. R. Svizero, R. C. B. Alonso, L. Wang, R. G. Palma-Dibb, M. T. Atta, P. H. P. D'Alpino. Kinetic of water diffusion and color stability of a resin composite as a function of the curing tip distance, *Mat. Res.*, **14**, 603-610 (2012).

[14]J. C. Pereira, P. H. P. D'Alpino, L. G. Lopes, E. B. Franco, R. F. L. Mondelli, J. B. Souza. Evaluation of internal adaptation of Class V resin composite restorations using three techniques of polymerization, *J. Appl. Oral Sci.*, **15**, 49-54 (2007).

[15]R. R. Reich, D. C. Sharpe, H. D. Anderson. Accelerated aging of packing: consideration, suggestions, and use in expiration date verification, *Med. Dev. Diag. Indus.*, **10**, 34-39 (1988).

[16]Y. Inaguma, C. Liquan, M. Itoh, T. Nakamura, T. Uchida, H. Ikuta, M. Wakihara, High ionic conductivity in lithium lanthanum titanate, *Solid State Commun.*, **86**, 689-693 (1993).

[17]C. Bonhke, B. Regrag, F. Le Berre, J. L. Fourquet, N. Randrianantoandro, Comparison of pH sensitivity of lithium lanthanum titanate obtained by sol–gel synthesis and solid state chemistry, *Solid State Ionics*, **176**, 73-80 (2005).

[18]A. Belous, O. Yanchevskiy, O. V'yunov, O. Bohnke, C. Bohnke, F. Le Berre, J. Fourquet, Peculiarities of $Li_{0.5}La_{0.5}TiO_3$ formation during the synthesis by solid-state reaction or precipitation from solutions, *Chem. Mater.*, **16**, 407-417 (2004).

[19]S. Scharner, W. Weppner, P. Schmid-Beurmann, Evidence of two-phase formation upon lithium insertion into the $Li_{1.33}Ti_{1.67}O_4$ spinel, *J. Electrochem. Soc.*, **146**, 857-861 (1999).

[20]M. Wagemaker, D. R. Simon, E. M. Kelder, J. Schoonman, C. Ringpfeil, U. Haake, D. Lutzenkirchen-Hecht, R. Frahm, F. M. Mulder, A kinetic two-phase and equilibrium solid solution in spinel $Li_{4+x}Ti_5O_{12}$, *Adv. Mater.*, **18**, 3169-3173 (2006).

[21]C. Legrand, J. Delville, Sur les paramètres cristallins du rutile et de l'anatase. *C. R. Hebd. Seances Acad. Sci*, **236**, 944-946 (1953).

[22]E. P. M. Meagher, G. A. Lager, Polyhedral thermal expansion in the TiO, polymorphs: Refinement of the crystal structures of rutile and brookite at high temperature, *The Canadian Mineralogist*, **17**, 77 (1979).

[23]B. L. Bischoff, M. A. Anderson, Peptization process in the Sol-Gel preparation of porous anatase (TiO_2), *Chem. Mater.*, **7**, 1772-1778 (1995).

CERAMIC GAS-SEPARATION MEMBRANES FOR ADVANCED ENERGY APPLICATIONS

C.A. Lewinsohn, J. Chen, D. M. Taylor
Ceramatec Inc.
2425 South 900 West
Salt Lake City, UT, 84119

P.A. Armstrong, L.L. Anderson, M.F. Carolan
Air Products and Chemicals, Inc.
7201 Hamilton Blvd.
Allentown, PA   18195-1501

Certain oxide materials conduct both electronic and ionic species and are called "mixed-conducting" materials. In particular, materials that conduct either oxygen or hydrogen can be used as gas separation membranes in a number of advanced energy applications. A general review of typical materials and the principles of using them for gas separation will be presented. Applications in gasification with carbon capture and storage, gas-to-liquids fuel processing, and hydrometallurgical refining will be reviewed. Engineering issues related to practical device design will be described. An overview and progress in current demonstration projects will also be presented.

INTRODUCTION

Transport of oxygen through dense layers of certain oxides with the fluorite structure, specifically zirconia, was shown to be practical for use in fuel cells in 1899[1]. It took almost 100 years, however, for Teraoka to discover that certain oxides with the perovskite structure exhibited orders of magnitude higher flux[2]. Even more recently, attention has focused on oxides exhibiting both ionic and electronic conductivity for transport of hydrogen[3]. In addition to industrial applications utilizing high volumes of gas, i.e. >1 000 l/h, materials that transport either oxygen or hydrogen, as a result of ionic conductivity, have applications in fuel cells, batteries, sensors, fluid delivery devices. Suitability for specific applications generally depends on the conductivity of the membrane, see Figure 1.

PASSIVE $< 100 \ \mu A/cm^2$	ACTIVE $< 10 \ mA/cm^2$	SUPERACTIVE $> 100 \ mA/cm^2$
• Sensors • Analysers • Micro-fluidic Delivery Devices	• DeOxo Systems • Batteries	• Industrial Membrane Reactor Systems • Fuel Cells

Figure 1   Suitable devices for materials exhibiting low, moderate, or high conductivity.

This paper will focus on materials and applications involving gas-separation membranes for industrial systems. In these applications, materials that have both ionic and electronic conductivity are desirable, since electroneutrality in the membrane can be maintained without the application of an external circuit. In simplified terms, application of a chemical gradient of the desired species provides a driving force for defect transport within the membrane. The chemical gradient can be established purely by absolute pressure gradients or by concentration gradients dictated by chemical equilibria, i.e. the equilibrium concentration of oxygen in a methane fuel environment. The defect flux is proportional to the mobility of the species, the concentration of the species, and the driving force[4]. At steady state, defect concentrations are determined by equilibrium equations at the membrane surfaces. Electroneutrality must be maintained. For example, combining the Wagner and Nernst-Einstein relationships, the flux of oxygen, j_{O2}, across an oxygen conducting membrane of thickness L due to a pressure gradient of oxygen is,

$$j_{O_2} \approx -\frac{D_V}{4V_m L}\int_{\ln P'_{O_2}}^{\ln P''_{O_2}} \delta\, d\ln p_{O_2}$$

Eq. 1

Where D_V is the oxygen vacancy diffusion coefficient, V_m is the molar volume of the oxygen vacancy, P'_{O2} is the partial pressure of oxide on one side of the membrane, P''_{O2} is the partial pressure of oxide on the other side of the membrane, and δ represents the concentration of oxygen vacancies. Detailed derivations of the governing equations are provided by Maier[5].

Materials

Oxides with the perovskite structures, i.e. $CaTiO_3$, are the most widely investigated and applied for industrial scale applications[6]. This is due to the values of ionic and electronic conductivity that can be obtained and their relative stability. Furthermore, perovskites are amenable to cation doping that can be used to enhance conductivity and stability[7]. Although perovskite structures are amenable to doping by a variety of elements and a wide range of dopant concentrations, the conductivity and physical properties of the resulting compounds is affected by the ionic radius and electronic structure of the dopants.

For applications involving the transport of oxygen, cobaltite-based perovskites and ferrite-based perovskites are the most commonly used. Cobaltite-based compounds typically can provide higher values of flux than ferrite-based ones[8], however their mechanical behavior is less robust. Strontium, lanthanum, and barium are the most commonly used A-site cations. $SrCo_{0.8}Fe_{0.2}O_3$ (SCF) has high oxygen transport[2], but at low temperatures it decomposes to a Brownmillerite phase with low conductivity due to oxygen vacancy ordering[9]. On the other hand, strontium doped, lanthanum cobaltite, $La_{1-x}Sr_xCoO_3$ (LSCO) exhibits a lower flux but greater phase stability. $La_{1-x}Sr_xCo_{1-y}Fe_yO_3$ (LSCF) has even better stability than LSCO, but lower flux. $Ba_{1-x}Sr_xCo_{1-y}Fe_yO_3$ (BSCF) exhibits both high flux and high phase stability, but is prone to decomposition in environments containing CO_2. Data reported[10] for the oxygen flux of these materials, for a constant membrane thickness, as a function of the gradient of the partial pressure of oxygen are show in Figure 2.

Similar to compositions that transport oxygen across a membrane via defect transport, there are also compositions that transport protons, which can be useful in applications requiring hydrogen transport. The principles of operation and governing equations are similar to those for oxygen-conducting membranes. Far fewer compositions with appreciable proton conductivity have been reported than with oxygen conductivity and there are some thermodynamic barriers to proton formation that present challenges. Nevertheless, due to the potential benefits in fuel processing and clean-power generation, research and development in this area continues. An excellent review of the atate of the art of dense, ceramic hydrogen membranes has been provided by Norby and Haugsrud[3].

As is the case with oxygen membranes, acceptor-doped perovskite materials are the leading candidates. Barium cerate, barium zirconate, and strontium cerate-based compounds are the most studied materials for application of dense, ceramic hydrogen conducting membranes[11,12,13]. BaPrO$_3$ and BaTbO$_3$ are other perovskite materials that have been investigated[14,15]. Although SrCeO$_3$ has approximately half the proton conductivity of BaCeO$_3$ materials, it has a lower oxygen ion-transport number and higher electronic transfer number than the BaCeO$_3$ materials[3]. In these materials, the low electronic conductivity limits the hydrogen flux. Efforts to improve the electronic conductivity of the cerates have focused on adding metallic and ceramic electronically conductive phases and various dopants[16].

In addition to the limitations on hydrogen flux imposed by low electronic conductivity, the cerate materials are also susceptible to reaction with CO$_2$ and phase instability in highly reducing conductions. Other materials that have been proposed as alternative, dense membranes are acceptor-doped rare-earth sesquioxides such as Tb$_2$O$_3$[17]. Significant proton conduction in materials with pyrochlore structures, Er$_2$Ti$_2$O$_7$[3] and La$_2$Zr$_2$O$_7$[18,19], has also been measured and La$_{0.9}$Ca$_{0.1}$NbO$_4$ was reported to have an extremely high proton conductivity as well as stability in CO$_2$ environments. Of these latter materials, only Er$_2$Ti$_2$O$_7$ has appreciable electronic conductivity, implying that external fields would be required to use the other materials as hydrogen pumps. La$_6$WO$_{12}$ has been reported to have both high proton conductivity[20] and considerable electronic conductivity[3] but it has unknown stability in environments relevant to operating conditions and, hence, is a target of continuing investigation.

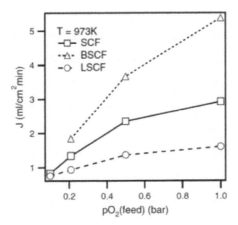

Figure 2   Oxygen flux, as a function of feed-side partial pressure of oxygen for SCF, BSCF, and LSCF[10].

Applications

One of the largest applications of oxygen is in steel making. Oxygen is used to produce carbon monoxide from coke, which is subsequently used to reduce iron from iron ore. Oxygen is also used to further remove carbon and excess silicon from iron while making steel. Another large scale application of oxygen is the combustion of hydrocarbon fuel during electricity generation or applications to control NOx emissions in high temperature processes, i.e. glass manufacturing. Oxygen is also used in hydro-metallurgical processes to obtain precious metals.

Oxygen also plays an important role in the conversion of gaseous fuel, such as natural gas, to liquid ones. Examples of the use of membrane systems utilizing the materials described above will be given in the following sections.

Typically during heat and power generation, fuel is combusted in air producing carbon dioxide, water, and nitrogen. To mitigate anthropomorphic climate change, carbon dioxide can be separated, or captured, from the combustion products and stored. This approach is known as carbon capture and storage (CCS). Part of the cost of capturing, transporting, and storing the carbon dioxide can be reduced by performing combustion in pure, or high concentrations of, oxygen. In this way, the combustion product stream is primarily carbon dioxide and water, which not only can be separated more easily, but also enables the lower costs associated with separating and handling additional species. Although combustion in pure oxygen, or high concentrations of oxygen, can offset some of the costs of CCS, there is an additional cost associated with producing large volumes of high purity oxygen. Therefore, there is great interest in utilizing the materials described in the previous section to reduce the cost of oxygen separation from air.

Since the flux of oxygen is thermally activated through the vacancy diffusion coefficient, i.e. Equation 1., practical values of flux through oxygen separation membrane materials usually occur at temperatures above 700-800°C. Therefore, dense, ceramic membranes are well suited for integration with other high temperature processes. An example of integration of an oxygen, ion transport membrane (ITM) system with electricity generation through gasification of coal or natural gas is shown in Figure 3.

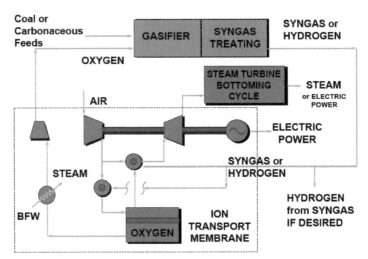

Figure 3   Block diagram illustrating integration of membrane-based, oxygen separation unit with an Integrated, Combined Cycle Gasification (IGCC) electricity generation plant[21].

Oxygen is also used in some mining operations to separate metals from their ores. For example, oxygen is used to oxidize iron is used in a high temperature, aqueous leaching process used to separate copper from chalcopyrite ore[22],

$$4\,Fe^{2+} + O_2 + 4\,H^+ = 4\,Fe^{3+} + 2\,H_2O \qquad\qquad \text{Eq. 2}$$

The oxidized iron is used to oxidize the chalcopyrite,

$$CuFeS_2 + 4\,Fe^{3+} = Cu^{2+} + 5\,Fe^{2+} + 2\,S \qquad\qquad \text{Eq. 3}$$

Giving a net reaction of,

$$CuFeS_2 + O_2 + 4\,H^+ = Cu^{2+} + 4\,Fe^{2+} + 2\,S + 2\,H_2O$$

The combination of lower cost, higher efficiency, and synergistic process integration provided by dense, ceramic membrane-based oxygen separation systems offer significant tangible and enabling benefits to many metals processing projects.

As mentioned above, oxygen is also used for the conversion of natural gas to liquid hydrocarbons. There are numerous pathways for converting natural gas to liquid, but in simplistic terms, methane can be partially oxidized, usually in the presence of steam, to form a mixture of hydrogen and carbon monoxide often referred to as synthesis gas or syngas. The syngas is then converted to a liquid through further reactions, such as the Fischer-Tropsch process,

$$(2n+1)\,H_2 + n\,CO = C_nH_{(2n+2)} + n\,H_2O \qquad\qquad \text{Eq. 4}$$

In addition to reducing the cost to produce large volumes of oxygen, dense, ceramic membrane systems also offer the potential for incorporation into membrane reactors where unit operations are combined in one reactor. For example, whereas a standard chemical processing facility may have a separate air separation unit (ASU) for separating oxygen from air, typically by cryogenic distillation, and a separate autothermal reformer for partially oxidizing methane to form syngas, both of these operations can be combined into a single unit operation using a membrane reactor.

Manufacturing

To obtain the benefits of dense, ceramic membrane materials in the application described above, components and devices must be fabricated via commercially viable methods. Supported by Air Products and Chemicals, Inc. and the U.S. Department of Energy, Ceramatec, Inc., has developed methods for large scale manufacturing of ceramic membrane components. For a variety of reasons, such as low pressure drop and mechanical reliability, Ceramatec has designed basic membrane components with a planar, microchannel architecture[23], see Figure 4. Although the examples used to illustrate the design and manufacturing are for oxygen separation from air, they are representative of modules made from other or similar materials, including hydrogen transport membrane materials, for different applications such as gas-to-liquids processing. In the planar component described, the dense, ceramic membrane is on the outer surface and it is supported internally by a porous support layer and microchannel layers that also serve to transport the permeate gas to the central manifold. Numerous individual

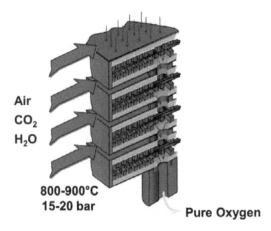

Figure 4   Schematic cross-section of planar, microchannel components and internal manifold for oxygen separation[21].

Figure 5   Modules of multiple microchannel components for oxygen separation.

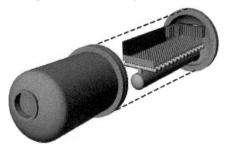

Figure 6   Schematic of a pressure vessel containing modules for oxygen production.

microchannel components are assembled into stacks, or modules, to obtain the desired capacity for oxygen separation, see Figure 5. Finally, for large scale industrial applications, numerous modules will be installed in a single, or multiple, pressure vessel, see Figure 6.

The planar microchannel components described are made by commercial, ceramic manufacturing processes. Ceramic powder, with the desired chemical and physical properties, is blended in solvent with organic binders, plasticisers and other additives to produce a slip suitable for tape casting. Various combinations of blade geometry, viscosity, and drying arrangements are used to cast rolls of tape ranging from one to several hundreds of microns thick. The tape can be featured by laser cutting or punching to form the regions that will become microchannels. The featured tape is laminated, using various combinations of heat, pressure, and solvents. The laminates are then sintered at elevated temperatures, to remove the organic material and cause the ceramic powder to densify.

The planar components can be assembled into stacks using a variety of sealants including ceramic inks, glasses, brazes, and diffusion bonding. To obtain the desired spacing for the mass and heat transport on the non-microchannel side of the system, auxiliary components may be required to obtain the desired configuration. Finally, manifold components can also be attached to enable connections with dissimilar materials, or to provide the desired connections to upstream and downstream components. Modules made by this approach have been tested successfully and have met flux and purity targets through a number of cycles in a $\approx$ 50 l/h prototype system[21].

SUMMARY

Numerous oxide materials exhibit oxygen or hydrogen transport properties that are attractive for industrial applications. Cobaltites with the Perovskite structure are the leading candidates for applications utilizing oxygen transport. Cerate-based Perovskites are the leading candidates for applications involving hydrogen transport, although the study of these materials is less mature than that of materials with oxygen transport, and several other materials are also under investigation. Dense, ceramic membranes offer the potential for reducing the cost of carbon capture and sequestration for large scale combustion processes such as electricity generation. Membrane materials also have applications in hydrometallurgical refining and gas-to-liquids fuel processing. In order to obtain the benefits provided by these materials, commercial methods for manufacturing membrane modules have been developed.

ACKNOWLEDGEMENTS

This report and the material herein are based upon work supported by the Department of Energy under Award Number DE-FC26-98FT40343.

Reference herein to any specific commercial product, process, or service by trade name, trademark, manufacturer, or otherwise does not necessarily constitute or imply its endorsement, recommendation, or favoring by the United States Government or any agency thereof.

The views and opinions of the authors expressed herein do not necessarily state or reflect those of the United States Government, any agency thereof, or Air Products and Chemicals, Inc. or it affiliates, or Ceramatec, Inc..

REFERENCES

[1] W. Nernst, "Reasoning of theoretical Chemistry: Nine Papers (1889-1921)," Frankfurt am Main : Verlag Harri Deutsch, (2003).

[2] Y. Teraoka, H. M. Zhang, S. Furukawa and N. Yamazoe, Chemistry Letters (1985) 1743.

[3] T. Norby, R. Haugsrud, "Dense Ceramic Membranes for Hydrogen Separation," Ch. 1 in Nonporous Inorganic Membranes, Sammells A.F. and Mundschau [ed.], Wiley-VCH Verlag GmbH & Co., 2006.

[4] C. Wagner, "Equations for Transport in Solid Oxides and Sulfides of Transition Metals," Prog. Solid State Chem., vol. 10, no. 1, pp. 3-16 (1975).

[5] J. Maier, "Mass transport in the Presence of Internal Defect Reactions – Concept of conservative Ensembles: I, Chemical Diffusion in Pure Compounds," J. Am. Ceram. Soc., vol. 76, no. 5, pp. 1212-1217 (1993).

[6] H.J.M. Bouwmeester, A.J. Burggraaf, "Dense ceramic membranes for oxygen separation," Membrane Science & Technology, vol. 4, pp.435-528 (1996).

[7] J.B. Goodenough, Rep. Prog. Phys., vol. 67, pp.1915 (2004).

[8] K. Zhang, J. Sunarso, Z. Shao, W. Zhou, C. Sun, S. Wang, S. Liu, RSC Advances, vol. 1, pp. 1661-1676 (2011).

[9] J.P. Hodges, S. Short, J.D. Jorgensen, J. Solid State Chem., vol. 151, pp. 190 (2000).

[10] J.F. Vente, S. McIntosh, W.G. Haije, H.W. Bouwmeester, J. Solid State Electrochem., vol. 10, pp. 581-588 (2006).

[11] H. Iwahara, T. Esaka, H. Uchida, N. Maeda, Solid State Ionics, vol. 3-4, pp. 359 (1981).

[12] H. Iwahara, T. Yajima, T. Hibino, K. Ozaki, H. Suzuki, Solid State Ionics, vol. 61, pp.65 (1993).

[13] A. Mitsui, M. Miyama, H. Yanagida, Solid State Ionics, vol. 22, pp. 213 (1987).

[14] T. Fukui, S. Ohara, S. Kawatsu, J. Power Sources, vol. 71, pp. 164 (1998).

[15] L. Li, J.R. Wu, M. Knight, S.M. Haile, Proc. Electrochem Soc. PV 2001-28, Ionic and Mixed Conducting Ceramics, pp. 58-66 (2002).

[16] S. Elangovan, B. Nair, T. Small, B. Heck, I. Bay, M. Timper, J. Hartvigsen, and M. Wilson, "Ceramic Membrane devices for Ultra-High Purity Hydrogen Production: Mixed Conducting Membrane Development," Ch. 4 in Inorganic Membranes for Energy and Fuel Applications, A.C. Bose [Ed.], Springer Science + Business Media LLC, 2009.

[17] R. Haugsrud, Y. Larring, T. Norby, Solid State Ionics, vol. 176, no. 39-40, pp. 2957-2961 (2005).

[18] T. Shimura, M. Komori, H. Iwahara, Solid State Ionics, vol. 86-88, pp. 685-689 (1996).

[19] J.A. Labrincha, J.R. Frade, F.M.B. Marques, Solid State Ionics, vol. 99, pp. 33 (1997).

[20] T. Shimura, S. Fujimoto, H. Iwahara, Solid State Ionics, vol. 143, pp. 117 (2001).

[21] Vratsanos L.A., Armstrong P.A., Underwood R.P., Stein V.E., Foster E.P., "Enabling Clean Energy Production From Coal: ITM Oxygen Development Update," Proceedings of the Int.l Pittsburgh Coal Conference, Pittsburgh, PA., September 2009.

22   Littlejohn, P.O.L. , Dixon D.G., "The Enhancing Effect of Pyrite on Ferrous Oxidation by Dissolved Oxygen," pp. 1098-1109 in Hydrometallurgy 2008: Proceedings of the Sixth International Symposium, C. Young, P. Taylor, C. Anderson, Y. Choi [Ed.], Society for Mining, Metallurgy, and Exploration, Englewood, CO (2008).

23   U.S Patent 5,681,373, 1997.

Advanced Materials and Technologies for Energy Generation and Rechargeable Energy Storage

Li-ION CONDUCTING SOLID ELECTROLYTES

Rost, A.(*); Schilm, J.; Kusnezoff, M.; Michaelis, A.

Fraunhofer Institute for Ceramic Technologies and Systems
Winterbergstr. 28
01277 Dresden
Germany

ABSTRACT

Solid electrolytes for Li-ion batteries offer large potential to obtain safe batteries and avoid internal short cuts as also high thermal stability. Suitable compositions are offered in the $Li_{1+x}Al_xTi_{2-x}(PO_4)_3$-system. Within this study glasses with varying TiO_2 to Li_2O and Al_2O_3 ratios were molten and crystallised by selected heat treatments and characterised. The crystalline phases are lithium analogues of the NaSICON-Structure. Impedance spectroscopic measurements were carried out for determining ionic conductivities with respect to the chemical compositions, crystallising conditions and temperature. The results prove pure ionic conductivity up to $1.6 \cdot 10^{-4}$ S/cm.

INTRODUCTION

In common state of the art Li-ion batteries, the both electrodes are kept in distance by porous polymeric separators which are filled with liquid, Li-ion conducting electrolyte. By adjusting an accurate distance between the electrodes, separators represent security relevant components. But in the case of warming up, the strength of polymers drops markedly and the distance between the electrodes can decrease what in turn leads to further increase of the batteries temperature. Hot spots can occur, leading to short circuits by direct contacts of the electrodes. The use of heat resistant ceramic powders as filler in porous, polymeric separators offers a possibility to achieve a higher thermal stability. A drawback of this solution is an increase of the internal ohmic resistance of the separators because most ceramics offer no conductivity for Li-ions [1; 2]. The use of ceramic materials with conductivity for Li-ions enable an additional contribution for the overall conductivity of the separator filled with liquid electrolyte and decrease the ohmic resistance of the batteries what in turn maintains a high level in operational safety. Furthermore ceramics can be produced as thin foils with adjustable porosity or even as dense substrates. Applied in this manner, Li-ion conductive ceramics can not only act as separators but moreover as solid electrolytes with various favourable advantages. Dense substrates can be applied as electrolytes in new battery concepts in which the anodic and cathodic half cells are separated hermetically. So the use of two different liquid electrolytes becomes possible. Furthermore the growth of Li-dendrites, what is a major degradation mechanism, can be stopped by a dense solid electrolyte. Finally solid and gastight electrolytes represent a key component of Li-air batteries which offer energy densities comparable to fossil liquid fuels.

The aim of this study focuses on synthesis and characterisation of Li-ion conducting glass ceramics in the Li-Al-Ti-P-O-system with high sintering ability [3 - 6]. A material based on the composition $Li_{1+x}Al_xTi_{2-x}(PO_4)_3$ with x=0.275, is reported to show high Li-ion conductivity [7].

To get dense and gastight solid electrolytes a good sintering ability is required. In Literature different details are available about compositions with high internal ionic conductivity and good sintering ability. Often it is stated that stoichiometric compositions show bad shrinkage and densification and so certain changes of the ideal composition $Li_{1+x}Al_xTi_{2-x}(PO_4)_3$ have to be performed to obtain a better sintering ability. So in this work the individual contents of Li and Ti are altered to examine the effects of sintering and ionic conductivity

EXPERIMENTAL

All solid electrolyte materials were manufactures by a typical glass processing route including subsequent milling and crystallisation processes. Li_2CO_3; $AlOH_3$; TiO_2 and $(NH_4)_2HPO_4$ were used as raw materials in reagent grade qualities. The batches were homogenised in a tumble mixer for 30 minutes and subsequently calcined by heating up at 1 K/min to 500 °C in alumina crucibles. For melting the mixtures were heated up to 1350 – 1430 °C and refined by annealing 1 h. The molten glasses were quenched by pouring them on a graphite base and after complete cooling the glasses were grinded to powders with $d_{50} = 6$ μm. Due to moderate cooling rates by pouring glass melts on graphite, in the cooling samples some crystallisation started and no complete glassy and amorphous samples were manufactured. This shows the difficulty in preparing Li-ion conducting separator materials and is in contrast to data reported in [7] by quenching on preheated steel plates.

However investigation of the sintering behaviour, hot stage microscopic analysis was carried out in a Leitz Hot Stage Microscope (optical dilatometer) by heating with 5 K/min up to 1000 °C in air. Further the powders were compacted to cylinders of 6 mm diameter and 10 mm length and sintered at temperatures between 600 and 1200 °C for up to 5 h to get solid and well crystallised test specimens for the analysis of ionic conductivity. For these measurements the plan-parallel abutting faces were sputtered with gold and pressed between two platinum electrodes. Impedance measurements at room temperature were carried out by a Gamry Reference 600 impedance analyser with an amplitude of the electric potential of ± 100 mV in a frequency range between 100,000 and 0.02 Hz

RESULTS AND DISCUSSION

To get dense and Lithium ion conducting solid electrolytes by sintering of glass or glass ceramic powders, compositions with high ionic conductivity and a good sinterability have to be found. Tabel 1 lists a choice of prepared compositions based on the stoichiometry $Li_{1+x}Al_xTi_{2-x}(PO_4)_3$. According to [7] the compositions A and B should show a good sintering and shrinking behaviour. In the compositions C and D the ratio of Li_2O and Al_2O_3 is increased further with respect to TiO_2.

Table 1: Molar ratios of the batch compositions

	A	B	C	D
Li_2O	14.00	16.25	20.00	22.50
Al_2O_3	9.00	3.75	7.50	10.00
TiO_2	38.00	42.50	35.00	30.00
P_2O_5	39.00	37.50	37.50	37.50
$Li^+ : Ti^{2+}$	1 : 2.7	1 : 2.3	1 : 1.75	1 : 1.3

The sintering ability can be deduced by the shrinkage behaviour of powder compacts measured by optical dilatometry in Fig. 1. The composition A shows nearly no shrinkage hence no densification up to a temperature of 1000 °C. According to first crystallisation during the quenching of the glass melt, the present crystals grow further and avoid any sintering shrinkage. In contrast the compositions B to D show increasing degrees of shrinkage and densification with respect to a decreasing Li_2O and Al_2O_3 to TiO_2 ratio. Also the in the same manner the onset of the shrinking process decreases from 865 °C for $Li^+ : Ti^{2+} = 1 : 2.3$ to 675 °C for $Li^+ : Ti^{2+} = 1 : 1.3$ and the maximum shrinkage increases from 2.1 % to 10.8 % linear shrinkage.

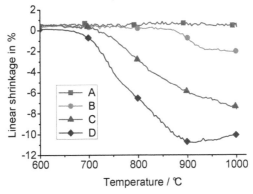

Fig. 1: Linear shrinkage detected by hot stage microscopy

The powder compacts made of composition A show virtually no changes of the initial green density. So it can be assumed, that the initial crystallising at quenching the molten glass progressed in a large extend, so that nearly no sinter ability by means of any densification remained.

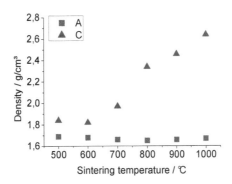

Fig. 2: density after heat treatment for 5 h at temperatures between 500 and 1000 °C

When composition A shows no detectable shrinkage during sintering up to 1000 °C, the density should not increase markedly. To examine this, density measurements after heat treatments up to 1000 °C for 5 h were carried out. Fig. 2 shows clearly that for composition A there is no increase of density in the relevant temperature range. As it is can be seen for the example of composition C the other three compositions instead show densifications. With sintering a densification takes place and the density is increased from about 1,8 g·cm^{-3} to 2,64 g·cm^{-3} when treated at 1000 °C.

A high ionic conductivity depends on the mix of the crystallised phases. The aim is to get a maximum amount of $Li_{1+x}Al_xTi_{2-x}(PO_4)_3$ and as less as possible of non conducting phases and also less glassy phase. From DTA investigations not shown here it is known that crystallisation processes start at temperatures as low as 500 °C what is well below the start of the sintering. Therefore XRD investigations were performed after heat treatments between 500 and 1000 °C (Fig. 3). As can be seen, even at 500 °C $Li_{1+x}Al_xTi_{2-x}(PO_4)_3$ can be detected. With increasing temperature also the intensitiy of this desired phase increases in all composition. In Addition always some minor phases are detectable. The development of tem, strongly depends of the individual composition. For composition A, $AlPO_4$ was detected at temperatures above 800 °C. Compistion C instetad shows Peak of an unkwon phases between 700 and 800 °C. For higher tempeatures this phaes dissolves. This different development of the phase composition shows that the crystallisation temperature for different composition has to be determined individually.

▼ $Li_{1+x}Al_xTi_{2-x}(PO_4)_3$
■ $AlPO_4$
• unknown

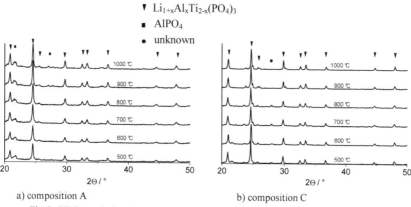

a) composition A   b) composition C

Fig 3: XRD analysis of composition A and C after distinct heat treatments

With respect to the different sintering abilities, impedance measurements were carried out after crystallising powder compacts at temperatures between 600 and 1200 °C (Fig. 3). As expected in accordance with the shrinking behaviour composition A has shows only a low ionic conductivity. After crystallization for 5 h at 600 °C the conductivity reaches only a low level of 3.4·10^{-8} S/cm. By increasing the crystallization temperature to 800 and 1000 °C the conductivity decreases by one order of magnitude down to 3.4·10^{-9} S/cm. This indicates that at higher temperatures no conducting sintering contacts but moreover isolating phases between the grains have been formed. This in turn results in continuously decreasing ionic conductivities.

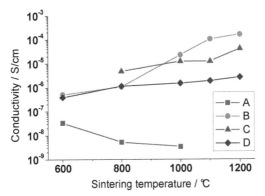

Fig. 4: Conductivity of solid electrolyte compositions after heat treatment at different temperatures. The lines are only guides for the eyes.

In case of the compositions B to D increasing sintering temperatures result in higher conductivities. The missing value for composition C at 600 °C is explained by a low strength of the powder compacts at this sintering temperature which did not allow preparing specimens for the measurement. As Fig. 3 shows, the individual conductivities cover a broad ranging from $4.0 \cdot 10^{-7}$ S/cm to $1{,}6 \cdot 10^{-4}$ S/cm. After crystallization at 600 °C and 800 °C the compositions B and D show nearly the same conductivities. For higher sintering temperatures the conductivities diverge strongly.

Compared to 800 °C the conductivity of composition D is doubled at 1200 °C and reaches $2.7 \cdot 10^{-6}$ S/cm. The composition C the shows similar characteristic but at a higher level so a maximum conductivity of $4.2 \cdot 10^{-5}$ S/cm was measured. In contrast, for composition B much higher conductivities can be reached. Compared to composition D at temperatures above 800 °C the conductivity is increased drastically to $1.6 \cdot 10^{-4}$ S/cm after sintering at 1200 °C.

It is remarkable, that not the compositions C and D with high sinter shrinkages lead to high ionic conductivities but composition B which offers despite of A the smallest densification. The more and more increasing conductivity for composition B at higher sintering temperatures in respect to composition D show that changes in microstructure must take place resulting in high conductivity for Li-ions. Two reasons can be mentioned therefore. One is a higher amount of conducting crystals. And the other is that the crystalline phase of composition B is more conductive as the one of composition D. As the amount of Al_2O_3 in composition D is nearly the threefold of the amount used in composition C, the first reason is not probably. In the structure of $Li_{1+x}Al_xTi_{2-x}(PO_4)_3$, not only the total ammount of Li is important, but also the mobility of the Li-ions. When a composition with relative low Al_2O_3-content gains higher condictivities in respect to compositions with higer contents of Al_2O_3 and also higer values of sintering shrinkage, the mobility of the Li-ions might be the determing factor for high conductivity. Nevertheless dense microstructures with low porosity and high crystalinity exibit higher conductivities and so the conductivity for all compositions increase with increasing sintering temperature.

CONCLUSION

In this study the sinter ability and conductivity of compositions of the system $Li_{1+x}Al_xTi_{2-x}(PO_4)_3$ was examined. Due to some crystallisation at quenching the glass melts, partially low sintering abilities and densifications followed. To get high ionic conductivity, not only dense material is important but also conductive transitions for Li-ions between the individual crystals. In the case of this work highest conductivities of $1.6 \cdot 10^{-4}$ S/cm occurred for a composition of $16.25 \cdot Li_2O\text{-}3.75 \cdot Al_2O_3\text{-}42.50 \cdot TiO_2\text{-}37.50 \cdot P_2O_5$ after sintering for 5 h at 1200 °C.

REFERENCES

[1] Ahmad, S., Agnihotry, S. A.: *Effect of nano γ-Al2O3 addition on ion dynamics in polymer electrolytes.* Current Applied Physics, 9, S. 108 - 114, 10-9-2007.

[2] Ahn, J.-H., Wang, G. X., Liu, H. K., Dou, S. X.: *Nanoparticle-dispersed PEO polymer electrolytes for Li batteries.* Journal of Power Sources, 119-121, S. 422 - 426, 2003.

[3] Fu, J.: *Superionic conductivity of glass-ceramics in the system Li2O--- Al2O3---TiO2--- P2O5.* Solid State Ionics, 96 [3-4], S. 195 - 200, 14-5-1996.

[4] Fu, J.: *Fast Li+ ion conducting glass-ceramics in the system Li2O-Al2O3-GeO2-P2O5.* Solid State Ionics, 104 [3-4], S. 191 - 194, 30-6-1997.

[5] Fu, J.: *Effects of M3+ Ions on the Conductivity of Glasses and Glass-Ceramics in the System Li2O—M2O3—GeO2—P2O5 (M = Al, Ga, Y, Dy, Gd, and La).* Journal of the American Ceramic Society, 83 [4], S. 1004 - 1006, 19-4-1999.

[6] Abrahams, I., Hadzifejzovic, E.: *Lithium ion conductivity and thermal behaviour of glasses and crystallised glasses in the system Li2O-Al2O3-TiO2-P2O5.* Solid State Ionics, 134 [3-4], S. 249 - 257, 8-5-2000.

[7] Thokchom, J. S., Kumar, B.: *Ionically Conducting Composite Membranes from the Li2O– Al2O3–TiO2–P2O5 Glass–Ceramic.* Journal of the American Ceramic Society, 90 [2], S. 462 - 466, 29-8-2006.

SODIUM IRON PHOSPHATE $Na_2FeP_2O_7$ GLASS-CERAMICS FOR SODIUM ION BATTERY

Tsuyoshi Honma, Takuya Togashi, Noriko Ito, Takayuki Komatsu
Department of Materials Science and Technology, Japan

ABSTRACT

New cathode candidate $Na_2FeP_2O_7$ for rechargeable sodium ion second battery was successfully prepared by glass-ceramics method. The precursor glass, which is same composition in $Na_2FeP_2O_7$, was prepared by melt-quenching method. $Na_2FeP_2O_7$ was obtained by heat treatment of precursor glass powder with glucose addition as reduction agent of iron oxide at $620°C$ for 3h in electric furnace. $Na_2FeP_2O_7$ has triclinic P1- structure. By means of electrochemical charge-discharge testing, $Na_2FeP_2O_7$ exhibits 2.9V, 88mAh/g, in which is 90% for the theoretical capacity during 2.0-3.8V cut-off voltages. $Na_2FeP_2O_7$ ceramics has the potential for the safety cathode candidate for the sodium ion battery with a low materials cost.

INTRODUCTION

Since Sony launched commercial lithium ion battery(LiB) in 1991, LiB has widely used in many kinds of electronic devices. In recently, LiB has expand from portable electronics to large scale applications, especially, electrical vehicles. Poly-anion based $LiFePO_4$ has been much attracted because of its remarkable electrochemical storage properties for the next generation LiB.[1] $LiFePO_4$ has superior thermal and electrochemical stability compared with conventional $LiCoO_2$, $LiMn_2O_4$ cathode materials. The capacity of battery package in electrical vehicle requires over than 10kWh, therefore pretty much lithium resources will be required. However, lithium resources are unevenly distributed in the world hence lithium metal is classified as rare-metal resource. In near future the problem concern about materials cost must be happening. On the other hand the other alkaline and alkaline earth ion based secondary batteries are proposed. In especially, sodium (Na) is located below Li in the periodic table and they share similar chemical properties in many aspects. The fundamental principles of the sodium ion battery(NaB) and LiB are identical; the chemical potential difference of the alkali-ion (Li or Na) between two electrodes (anode and cathode) creates a voltage on the cell. Some articles concern about cathode materials are reported so far. Typically, layered rock salt type α-$NaCrO_2$, which is same as $LiCoO_2$, is known as cathode active materials for sodium ion battery.[2] Although the layered rock salt exhibits good electronic conductivity and sodium ion intercalation, but the use of rare metals prohibits its utilization.

Figure 1 shows the glass-formation region in the Na$_2$O-Fe$_2$O$_3$-P$_2$O$_5$ ternary glass system. In Na$_2$O-Fe$_2$O$_3$-P$_2$O$_5$ system there are four NaB cathode candidates ever reported.

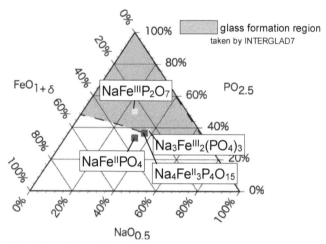

Figure 1. Glass formation region of Na$_2$O-Fe$_2$O$_3$-P$_2$O$_5$ system. The candidates of cathode material are also shown.

Phosphate based maricite NaFePO$_4$, which is same composition in LiFePO$_4$, is not suitable for the cathode materials because of complex poly anion units disturbing sodium ion conduction.[3] Therefore there are only a few polyanions know to be candidates as cathode. Development of new cathode active materials is needed for the realization of sodium ion battery. By using crystallization process of glass materials, it is easy control crystal morphology and size distribution from uniform glass matrix. We are proposing unique technique to fabricate phosphate based cathode materials such as olivine structured LiFePO$_4$ and NASICON structured Li$_3$V$_2$(PO$_4$)$_3$ ceramics by crystallization process from glass precursor.[4-8] Precursor glass has very homogeneous compositional distribution hence we obtained any ferromagnetic byproduct free LiFePO$_4$ which works well even in the high current density condition. We recently report the new cathode candidate Na$_2$FeP$_2$O$_7$ fabricated by glass-ceramics method at first. In this study we report the crystallization phenomena and electrochemical properties in Na$_2$FeP$_2$O$_7$.

EXPERIMENTAL

The Na$_2$FeP$_2$O$_7$ precursor glass was fabricated by conventional melt-quenching method.

Starting reagent NaPO$_3$ (Nakarai tesque) and Fe$_2$O$_3$ (Kojyundo chemicals) was mixed and melted in platinum crucible at 1100°C for 10min in electric furnace. By pouring melts on the steel plate the black-colored precursor glass was successfully formed. The glass transition temperature and crystallization temperature was determined by differential thermal analysis (DTA, Rigaku TG-8120). Glass powder which grain size is about 2μm was obtained by using of planetary ball mill (Fritsch premium line P-7). Glass ceramics was prepared by heat treatment in 5%H$_2$-95%Ar gas flowed tubular electric furnace at 620°C for 3h. To reduce Fe^{3+} ion in precursor glass 10wt% glucose was added glass powder. The amount of residual carbon content was determined by thermogravimetric analysis (TG-DTA, Rigaku TG-8120). Powder X-ray diffraction (XRD, Rigaku UltimaIV) employing Cu Kα radiation was used to identify the crystalline phase of prepared powders. The concentration of Fe^{2+} in the glass and glass-ceramics with the composition of LiFePO$_4$ was determined using a cerium redox titration method, in which 0.1N-Ce(SO$_4$)$_2$/H$_2$SO$_4$ aqueous solution as titrant and ortho-phenanthoroline as indicator were used. The cathode electrodes were fabricated from a mixture of active material, polyvinylidene fluoride (PVDF) and conductive carbon black in a weight ratio of 85:5:10. N-Methyl pyrrolidone (NMP) was used to make the slurry of the mixture. After homegenenization, the slurry was coated on a thin aluminum foil and dried at 90°C for 10h in a vacuum oven. The electrode was then pressed and disks were punched out as 16mmφ. The electrochemical cells were prepared using coin type cells. Sodium metal foils were used as anode, and glass filter paper (Advantec Co., GA-100) was used as separator. Test cell was assembled in an argon-filled glove box. The dew point of Ar atmosphere in glove box was kept as -86°C. The solution of 1M NaPF$_6$ (Tokyo Kasei Co.) in a mixture of ethylene carbonate (EC) and diethyl carbonate (DEC) (1:1, v/v, Kishida Chemicals Co.) was used as electrolyte. The cells were examined by using a battery testing system (Hokuto-denko Co.) at current density of 1/10 (0.02 mA/cm^2) for the theoretical capacity as 97mAh/g between 2.0 and 3.8 V.

RESULTS AND DISCUSSION

The bulk DTA pattern of Na$_2$FeP$_2$O$_7$ precursor glass is shown in Fig. 1. The glass transition (T_g) was determined as 451°C and the crystallization peak was determined as 580°C respectively. Furthermore an endothermic temperature was found at 693°C that corresponding melting point.

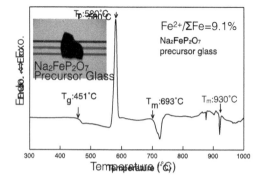

Figure 2. Differential thermal analysis (DTA) curve in Na$_2$FeP$_2$O$_7$ precursor glass.

According from the results of DTA measurement we determined heat-treatment temperature to fabricate Na$_2$FeP$_2$O$_7$ glass-ceramics as 620°C. The powder XRD patterns for the as quenched precursor glass and glass-ceramics heat-treated at 620°C for 3h in H$_2$-Ar gas were shown in figure 3.

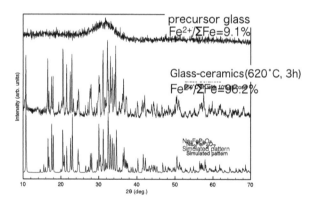

Figure 3. The powder XRD patterns for the as quenched precursor glass, glass-ceramics and simulated patterns of Na$_2$FeP$_2$O$_7$.

The percentage of residual carbon content was estimated as 1.9 wt% by means of TG-DTA analyses. It clearly seems sharp diffractions corresponding to the presence of crystalline phase. To refine the crystal structure in Na$_2$FeP$_2$O$_7$ we examined the Rietvelt analysis. We used and modified the crystal structure of triclinic Na$_{3.12}$Fe$_{2.44}$(P$_2$O$_7$)$_2$.[8] The determined crystal structure of Na$_2$FeP$_2$O$_7$ was illustrated in figure. 4 and the simulated XRD pattern is also shown in figure 3. All diffractions are agreed with simulated patterns. Obtained Na$_2$FeP$_2$O$_7$ has triclinic $P1$-structure. The parameters of unit cell are a = 0.64061nm, b=0.938893 nm, c=1.09716nm, α = 64.5381°, β= 86.0580°, γ=73.0619°. The presence of mainframe consisted from P$_2$O$_7$ and FeO$_6$ unit is a unique feature.

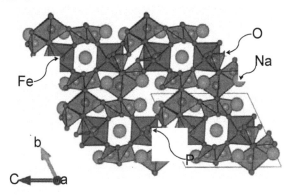

Figure 4. Crystal structure of triclinic Na$_2$FeP$_2$O$_7$.

According to crystal structure shown in figure 4 large tunnel structure is exist along with (100) direction to diffuse sodium ions. The precursor glass melted under air condition. According to the red-ox titration to determine valence state of iron, the precursor glass contains 9.1% Fe^{2+} for the total amount of Fe and the glass-ceramics contains 96.2% Fe^{2+}. It is clear that valence state completely change from Fe^{3+} to Fe^{2+} during heat-treatment. Hence we obtained single phase Na$_2$FeP$_2$O$_7$ from precursor glass. We also confirmed in previous study in the case of LiFePO$_4$ olivine phase from Fe^{3+} rich precursor glass. We examine the cathode properties of Na$_2$FeP$_2$O$_7$ in sodium ion battery. The electrochemical reaction of Na$_2$FeP$_2$O$_7$ crystal is expressed as following equation

$$Na_2Fe^{(II)}P_2O_7 \rightarrow NaFe^{(III)}P_2O_7 + Na^+ + e^- \qquad (1).$$

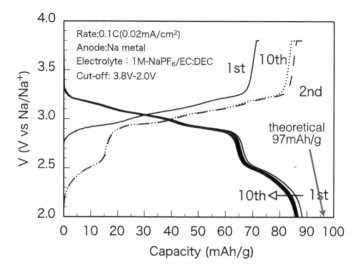

Figure 5. Charge-discharge curves during 10 cycles in test cell. The discharge rate was fixed as 0.1C

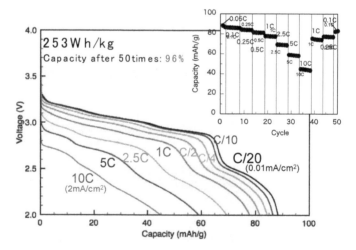

Figure 6. Discharge profile from 0.05C to 10C, size distribution of glass-ceramics/carbon powder and HR-TEM image around surface in $Na_2FeP_2O_7$/C glass-ceramics.

The theoretical capacity of Eq (1) is estimated as 97 mAh/g. The charge-discharge curves during 10 cycles in 0.1C are shown in Fig. 4. The initial discharge capacity is obtained as 88 mAh/g, which is corresponding to 90% for theoretical capacity. After 10 cycles the discharge capacity kept as 85mAh/g. It is considered that reversible electro-chemical reaction is available as shown in Eq(1). The principal plateau voltage exists at 2.9V and another plateau was partially observed at 2.5V. It seems that sodium ion site is classified as least two different sites as shown in Fig.3 however the ionic conduction mechanism is not cleared. The principal sodium ion conduction in $Na_2FeP_2O_7$ crystal might be dominated by Na(1) which is exists opened tunnel structure in crystal. On the other hand Na(2) exists complicated sites in poly-anion frame structure. We are going to clear the sodium ion conduction in near future.

CONCLUSION

We fabricated $Na_2FeP_2O_7$ triclinic phase which working as cathode active materials in sodium ion battery by glass-ceramics method successfully. It cleared that $Na_2FeP_2O_7$ glass-ceramics shows the good electrochemical properties, which is almost same as theoretical discharge capacity with a good cycle performance. Although the energy density is lower than that of $LiFePO_4$, $Na_2FeP_2O_7$ glass-ceramics has huge advantages for the reducing materials cost.

ACKNOWLEDGEMENT

This work was supported by the Grant-in-Aid for Scientific Research from the Ministry of Education, Science, Sport, Culture, and Technology, Japan (No. 23246114, 23655194 and 24656379), and partly by Program for High Reliable Materials Design and Manufacturing in Nagaoka University of Technology. One of authors (T. Honma) was financial supported from Ohkura-Kazuchika foundation in this study. The author would like to thank prof. T Kobayashi for the grain size distribution measurements.

REFERENCES
[1] A.K. Padhi, K.S. Nanjundaswamy and J.B. Goodenough, Effect of Structure on the Fe^{3+}/Fe^{2+} Redox Couple in Iron Phosphates, J. Electrochem. Soc., **144** 1609-13 (1997).

[2] S. Komaba, T. Nakayama, A. Ogata, T. Shimizu, C. Takei, S Takada, A Hokura, I Nakai, Electrochemically Reversible Sodium Intercalation of Layered $NaNi_{0.5}Mn_{0.5}O_2$ and $NaCrO_2$, ECS Trans. **16 (42)**, 43-55 (2009).

[3] C. M. Burba, R. Frech, Vibrational spectroscopic investigation of structurally-related $LiFePO_4$, $NaFePO_4$, and $FePO_4$ compounds, Spectrochimica Acta Part A, **65(1)** 44-50 (2006).

[4] K. Hirose, T. Honma, Y. Benino, T. Komatsu, Glass–ceramics with LiFePO4 crystals and crystal line patterning in glass by YAG laser irradiation, Solid State Ionics, **178 (11-12)**, 801-807 (2007).

[5] T. Honma, K. Hirose, T. Komatsu, T. Sato, S. Marukane, Fabrication of LiFePO4/carbon composites by glass powder crystallization processing and their battery performance, Journal of Non-Crystalline Solids, **356 (52-54)**, 3032-36 (2010).

[6] K. Nagamine, T. Honma, T. Komatsu, A fast synthesis of $Li_3V_2(PO_4)_3$ crystals via glass-ceramic processing and their battery performance, Journal of Power Sources, **196(22)**, 9618-24 (2012).

[7] T.Nagakane, H. Yamauchi, K. Yuki, M. Ohji, A. Sakamoto,T. Komatsu, T. Honma, M. Zou, G. Park, T.Sakai, Glass-ceramic LiFePO4 for lithium-ion rechargeable battery, Solid State Ionics, **206(5)**, 78-83 (2012).

[8] T. Honma, T. Togashi, N. Ito, T. Komatsu, Fabrication of $Na_2FeP_2O_7$ glass-ceramics for sodium ion battery, Journal of the Ceramic Society of Japan, **120(1404)**, 344–346 (2012).

HETEROGENEOUS MANGANESE OXIDE-ENCASED CARBON NANOCOMPOSITE FIBERS FOR HIGH PERFORMANCE PSEUDOCAPACITORS

Qiang Li[1,2], Karen Lozano[2], Yinong Lü[3], Yuanbing Mao[1,*]
[1]Department of Chemistry, University of Texas - Pan American, Edinburg, TX 78539 USA
[2]Department of Mechanical Engineering, University of Texas - Pan American, Edinburg, TX 78539 USA
[3]Department of Materials Science and Engineering, Nanjing University of Technology, Nanjing, Jiangsu 210009 China

ABSTRACT
    The integration of transition metal oxide nanocrystals and one-dimensional (1D) conducting carbon structures to generate their hybrids can create unpredictable new physical and chemical properties in comparison with single phase components. Here we report the fabrication of heterogeneous MnO nanocrystal (NC)-encased hierarchical carbon nanocomposite fibers (MCNFs) via a novel and large-scale Forcespinning followed by low temperature carbonization. Manganese nitrate containing polyvinylpyrrolidone (PVP) polymeric fiber was carbonized at relatively low temperature, $i.e.$ 500 °C, due to oxidant Mn^{2+} cations. Different inward and outward ionic diffusion rates of Mn^{2+} cations concurrently result in congregating MnO NCs near the surface region of the nanocomposites during thermolysis. After anodic and cyclic voltammetric electrochemical oxidations, in situ phase transformation from electrochemically inactive MnO NCs to pseudocapacitive MnO_x counterparts occurs, which yields a MnO_x NC/carbon hybrid fiber network with MnO_x NC-enriched functional surface. These NCs are accessible to aqueous electrolyte ions for Faradic redox reactions. Therefore these unique nanocomposites demonstrate a promising potential as pseudocapacitive electrode materials.

INTRODUCTION
    The integration of transition metal oxide nanocrystals (TMONCs) and one-dimensional (1D) carbon based skeletons has been exploited for various applications, such as chemical sensors,[1-2] catalysis,[3] nanoelectronics[4] and electrochemical devices.[5-7] In view of the united advantages of functionality of oxide nanocrystals and superior physical characteristics including electrical conductivity, mechanical tolerance, thermal stability and surface area of carbonaceous nanostructures, rationally constructing TMONC/carbon hybrids is prone to achieve unprecedented physical and chemical features with respect to single component counterparts.[8-9]
    Particularly for supercapacitor applications, two well-established synthetic strategies have been widely carried out to fabricate 1D TMONC/carbon hybrids: (1) pseudocapacitive TMONCs are anchored onto conductive carbon-based backbones, such as single/multiple walled carbon nanotubes, 1D conducting polymers and carbon cloth, via a variety of approaches including hydrothermal method,[10] electrochemical depositions,[6] atomic layer deposition,[11] precursor hydrolysis,[12] and so forth;[5] and (2) 1D TMO nanowires/nanoribbons with high aspect ratio are surface-shielded by conducting polymers through monomer polymerizations.[7, 13] For the first protocol, hybrid electrodes would suffer from inevitably uncontrolled ionic and electronic diffusions due to fast growth of active oxides during anchoring, capacitance fading caused by unexpected reactions between electrode materials and electrolytes as well as dissolution of active materials into electrolyte solutions, and pulverization problem because of particle aggregation during charge and discharge cycles. On the other hand, the second protocol is unlikely to offer

stable and fast paths for ionic diffusion and electronic transportation due to the intrinsic instability of conductive polymers and relatively large size of active materials. Even though in situ carbonation of electrospun hybrid polymeric fibers can incorporate nanocrystals into conducting carbon frameworks, the pseudocapacitive target oxides would be reduced into electrochemically inactive metal or metal oxides with lower oxidation states of transition metals during high temperature carbonations and the nature of homogeneously dispersed NCs in carbon matrix can deteriorate the rate capacity when limited electrolyte ion penetration depth is taken into consideration. Those are the primary reasons why carbonized hybrids have been explored mostly as anode materials with enhanced stability of lithium ion batteries.[14-16] Therefore, establishing an efficient, facile and productive path to fabricate Faradic TMONCs into carbon frameworks is still a tremendous challenge for excellent recyclability, high rate pseudocapacitors.

In this work, we present a novel strategy for heterogeneous MnO_x NCs/carbon nanocomposite fibers via high-yield Forcespinning® (FS) followed by carbonization in inert atmosphere and combined with electrochemical oxidations by applying three-electrode configuration. Acting as an updated substitution of electrospinning, FS eliminates the necessity of taking solution/melt electrical conductivity, electrical field strength, surface charge and ionization field in account by simply employing high speed centrifugal force and thus significantly simplifies the fabrication process of fine nanofibers.[17-18] Moreover, the high production rate, ~1g/min, affords great opportunities to introduce nano-scaled 1D materials into commercial manufacturing in various areas. In this case, $Mn(NO_3)_2$/PVP fibers (MPFs) were spun onto current collectors directly without adding any polymer binder and conductive additive for the first time to form intercross networks for facile electrolyte penetration and boosted gravimetric capacitance. Subsequent calcinations led to chronologically uneven ionic diffusion of Mn^{2+} ions, thermolysis and carbonization for forming hierarchical hybrid fibers. These fibers consist of MnO NCs evenly distributed on the near surface regions but simultaneously shielded by conducting carbon. The MnO NC/carbon hybrid carrying electrodes were further phase transformed into pseudocapacitively promising MnO_x NC/carbon hybrid fibers through anodic and cyclic voltammetric electrochemical oxidations. Remarkably, the electrochemical evaluation revealed that the final nanocomposite electrodes possess a maximum specific capacitance of 256.4 F g^{-1} when normalized to MnO_x in 1 M Na_2SO_4 electrolyte solution at a current density of 0.2 A g^{-1}, almost two orders of magnitude larger than the capacitance prior to oxidations, and superior rate performance.

EXPERIMENTAL

Polyvinylpyrrolidone (PVP, average MW = 130,000) and $Mn(NO_3)_2 \cdot xH_2O$ (99.99%, $x = 4$-6) were purchased from Sigma Aldrich. All reagents were of analytical grade and were used without further purification. The $Mn(NO_3)_2 \cdot xH_2O$/PVP fibers (MPFs) were prepared as follow. PVP (3.7 g) was gradually dissolved into 8.7 mL of 18.7, 10, and 25.6 wt.% $Mn(NO_3)_2 \cdot xH_2O$ aqueous solution by using both vortex mixing and sonication in 2-hour intervals for at least 48 h for a homogeneous dissolution. The precursor solution was maintained statically in dark in a vacuum oven at room temperature overnight to remove possible trapped air bubbles. To make MPFs, the prepared spinning solutions were fed into proprietary designed spinneret with evenly separated eight needles around its periphery. These needles have 0.29 mm inner orifice size and are 13 mm long. FS was carried out at a rotational speed of 9,000 rpm for 30 s, a 6 cm needle-to-collector distance with aluminum foil as collectors. The production rate in a lab scale unit was >1 g/min. Moreover, titanium substrates (0.25 mm thick, Aldrich), which were degreased ultrasonically prior to collection in acetone and ethanol for 10 min, respectively, were directly applied as current collectors. To directly deposit MPFs onto Ti substrate, Ti foil cut into 2 × 0.5 cm was fastened onto alumina collectors with polytetrafluoroethylene (PTFE) tape prior to the FS process. The MPFs were then calcined at 500 °C for 3 h in argon atmosphere with a heating ramp rate of 2 °C/min. These calcined MPFs, namely manganese monoxide MnO/C nanocomposite fibers (MCNFs), were oxidized using a Gamry reference 600 Potentiostat/ Galvanostat/ZRA workstation in a three-electrode cell system, in which 1 M Na_2SO_4 aqueous solution acted as electrolyte, and platinum gauze and Ag/AgCl were used as counter and reference electrodes, respectively. More specifically, a constant anodic oxidation was initially applied to the MCNFs at a current density of 10 μA cm^{-1} with a potential window from -0.6 to 0.9 V (vs. Ag/AgCl), and then a further oxidation was achieved after subsequent 1000 cyclic voltammetric scans between 0 and 1 V at 50 mV s^{-1}. After the oxidation processes, MCNFs converted into the final MnO_x NC-encased carbon nanocomposite fibers (EO-MCNFs).

CHARACTERIZATION

Field-emission scanning electron microscope (FESEM) characterizations were carried out on Carl Zeiss Sigma VP at 2 or 7.5 kV, equipped with backscatter electron detector (BSD). High resolution transmission electron microscopy (HRTEM) characterization was conducted on JEOL JEM-2010UHR coupled with selected area electron diffraction (SAED). The charging effect during SEM imaging on precursor MPFs was eliminated by coating with an approximately 100 Å thick Au/Pd layer using a Denton Desk II TSC turbo-pumped sputter coater. Fourier transform infrared (FTIR) spectra were recorded from 4000 to 450 cm^{-1} with a 4 cm^{-1} spectral resolution on a Thermal Nicolet Nexus 470 spectrometer with a DTGS detector by signal-averaging 32 scans. Differential scanning calorimetry (DSC) of the MPFs was run between 25 °C and 300 °C at a heating rate of 5 °C/min on a TA Instruments Q100 DSC in nitrogen environment. Thermogravimetic Analysis (TGA) was performed from 25 °C to 800 °C at 5 °C/min on a TA Instruments Q500 Thermogravimetric Analyzer for the MPFs and corresponding calcined MCNFs in nitrogen and air fluxes, respectively. The structural information of the calcined MCNFs were determined by X-ray powder diffraction (XRD) on a Rigaku Miniflex II with Cu Kα radiation (λ = 1.5418 Å) between 5° and 80°.

ELECTROCHEMICAL EVALUATIONS

All the electrochemical measurements were also conducted using the Gamry reference 600 Potentiostat/Galvanostat/ZRA workstation combined with PWR 800 software suit. Cyclic voltammetry (CV) were performed at a potential window of 0 to 1.0 V (vs. Ag/AgCl) with scan rates ranging from 2 mV s^{-1} to 100 mV s^{-1}. Galvanostatic charge/discharge testing was conducted between 0 and 1.0 V (vs. Ag/AgCl) at current densities ranging from 0.2 to 1.0 A g^{-1}.

RESULTS AND DISCUSSION

Figure 1 shows the SEM images of MPFs with low and high magnifications. MPFs were successfully spun into interconnected networks with no beads exhibiting via the FS method for different manganese nitrate contents. Bead-free feature can improve gravimetric performances when subsequently MPFs thermally decompose and phase transform into pseudocapacitive nanohybrids. For convenience, these MPFs are denoted as MPF-1, MPF-2 and MPF-3 with the increasing concentration of manganese nitrate. With increasing ratio of manganese nitrate to PVP, the as-spun MPFs exhibit more wrinkled surfaces. Additionally, these relatively randomly oriented MPFs form an intercrossing network. The average diameters of these MPFs vary from 400 nm to 1 μm for MPF-1, MPF-2 and MPF-3. According to previous FS studies, the fiber diameter is determined by multiple factors, such as spinning solution viscosity, solution surface tension, spinning rotational speed and spinneret needle to fiber collector distance.[17-18] Owing to the increases in $Mn(NO_3)_2 \cdot xH_2O$ concentrations, the precursor solution viscosity increases, which is mainly responsible for the variations of diameters of the MPFs.[19-20] The rotational force and polymeric wettability are creatively utilized to cement MPFs onto current collectors without introducing any polymer binder for electrode fabrications. Particularly noteworthy is that elimination of inactive binder and porous feature of network are critical to constructions of high rate devices in energy storage.

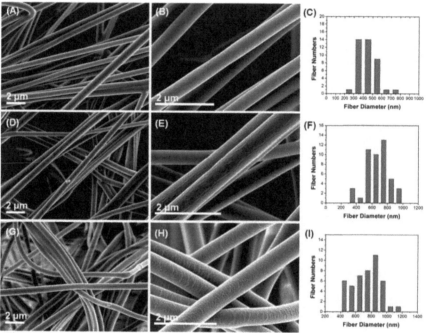

Figure 1. SEM images and diameter distributions of MPFs with different $Mn(NO_3)_2 \cdot xH_2O$ concentrations. (A, B, C) MPF-1, (D, E, F) MPF-2, and (G, H, I) MPF-3.

Figure 2 shows the FTIR spectra from the three MPF samples. The sharp absorption peak at 1383 cm^{-1} is attributed to N-O asymmetric stretching vibrations from NO^{3-} group.[21] Two remarkable broad peaks originating from 3400 to 3480 cm^{-1} and from 1668 to 1630 cm^{-1} correspond to O-H stretching and bending bands, respectively. They are derived from the residual water solvent, the hydrate water from manganese nitrate and absorbed moisture on the hygroscopic manganese nitrate and PVP. The fingerprint absorption modes of PVP are distinctly identified with prominent peaks at about 1277-1294, 1321, 1424-1468, 1500 cm^{-1} and 2916-2960 cm^{-1}, which are assigned to N-C stretching of $N-CH_2$, $C-H_2$ wagging, $C-H_2$ scissoring, N-C stretching of N-C=O, and symmetric C-H stretching, respectively.[22-23] The primary absorption mode at 1653 cm^{-1} corresponding to the carbonyl group C=O stretching vibration of PVP is probably superimposed with strong O-H bending vibration. The FTIR analyses indicate that the MPFs consist of nitrate anions and PVP polymer as well as water molecules. Along with the increasing manganese nitrate, the intensities of N-O stretching vibrations are apparently increased.

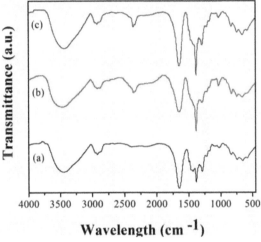

Figure 2. FTIR spectra of (a) MPF-1, (b) MPF-2, and (c) MPF-3.

Thermal experiments of the as-spun MPFs as well as PVP fibers were performed by DSC under nitrogen atmosphere, as showed in Figure 3. Each DSC thermogram from the three MPFs exhibits a distinct exothermic peak and an endothermic peak, even though the peak positions are shifting depending on different $Mn(NO_3)_2 \cdot xH_2O$ content. The endothermic peak located in relatively low temperature region is associated with the dehydration of PVP in the MPFs.[24] The upshifted degradation temperature is presumably owing to increasing $Mn(NO_3)_2 \cdot xH_2O$ content as well as the interactions between $Mn(NO_3)_2 \cdot xH_2O$ and PVP.[25-27] The presence of manganese nitrate gives rise to steric hindrance to the mobility of polymeric PVP backbone. Coordination complexes formed through van der Waals forces, polar attraction and stabilization from p-bond overlap reduce the polymeric PVP flexibility and dehydration rate while improve its thermal stability in the lower temperature region. On the other hand, the exothermic peaks that appear in the region of 210-283 °C is downshifting with increasing $Mn(NO_3)_2 \cdot xH_2O$ content. According to the XPS study toward the residual product from PVP degradation at different temperatures by Chen et al.,[28] the exothermic peaks can be attributed to the breaking of the C-N bonds linking pyrrolidone rings to the polymeric backbone during the decomposition of PVP. The premature degradation of PVP as shown by the downshifted exothermic peak with increasing $Mn(NO_3)_2 \cdot xH_2O$ content possibly results from the interaction between Mn^{2+} ions and carbonyl groups of PVP. This interaction hinders the formation of free radicals and subsequently that of their intermolecular termination reactions and thus accelerates the degradation of PVP.[14] Furthermore, Mn^{2+} ions can act as an oxidant to lower the minimum oxidation temperature of PVP.[29-30] Similar behaviors have been reported in literatures from nanocomposites. In addition, the heat released from the decomposition of PVP in the MPF-2 is much sharper than that from the other two MPFs. As it is well-known, thermogram study can be interpreted to structural variations. The degradation of PVP contains multiple stages, such as chain scission, cross-linking, and side-chain cyclization. The intensified peak may be affected by complex combination of thermally exothermic processes.[14]

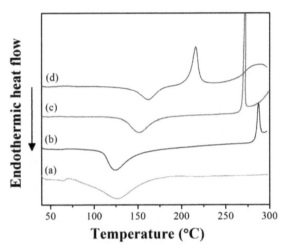

Figure 3. DSC thermograms of (a) pure PVP NFs, (b) MPF-1, (c) MPF-2, and (d) MPF-3 in nitrogen atmosphere.

Figure 4A shows the TGA thermograms of the MPFs and PVP fibers under nitrogen atmosphere. The first weight loss occurring around 50 – 110 °C is attributed to the evaporation of physically adsorbed water molecules. After that, the thermal degradation curves of MPFs show two decays whereas pure PVP fibers show only one decay, which is in accordance with other observations.[31] The degradation with the onset of 250 °C correlates with the exothermic peaks in DSC curves and corresponds to the breakage of pyrrolidone pendant groups from the PVP backbone. The second weight drop stretching up to 480 °C is attributed to the degradation of the main hydrocarbon chains of PVP. The differential of the weight loss (Figure 4B) highlights the two stages of PVP degradation more clearly. Interestingly no weight loss related to decomposition of anhydrous manganese nitrate from the MPNFs is distinguished in the range of 200 – 230 °C in DTG associated with DSC pattern, unlike thermal decomposition behavior of the pristine manganese nitrate. It is inferred that $Mn(NO_3)_2$ decomposition is postponed and occurs accompanying the degradation of the side chain of PVP due to the interaction between the manganese nitrate and PVP. The interaction may increase the activation energy of the formation of manganese oxide from manganese nitrate in PVP. The simultaneous degradation of manganese nitrate and PVP in the MPFs is supported by the fact that the decomposition percentage in the range of 263 – 315 °C increases proportionally with increasing the manganese nitrate content. Another interesting fact is that there is a slow but slight weight loss (~2 %) taking place after 480 °C from all three MPFs, in contrast to the pure PVP fibers. The decomposition of pure PVP completes at 480 °C. Thereby, the 2 % weight loss after 480 °C is tentatively ascribed to the reduction of MnO_2, the direct decomposition product of $Mn(NO_3)_2$, to lower valent manganese oxide by surrounding reductive carbon or carbon-rich compounds generated during carbonization.[15] In addition, about 18.1, 19.7, 22.6 and 3.4 wt.% residuals are eventually present from the calcination to 800 °C in nitrogen of the MPF-1, -2, -3 and pure PVP fibers, respectively. The higher residual percentage from MPFs than that from pure PVP fibers indicates the formation of manganese oxide/carbon composites.

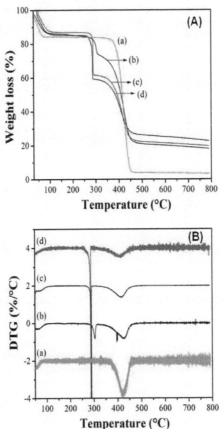

Figure 4. (A) TGA thermograms and (B) related DTG of (a) pure PVP fibers, (b) MPF-1, (c) MPF-2, and (d) MPF-3 in nitrogen atmosphere.

MPFs are thermally transformed into heterogeneous MnO NC-encased carbon nanocomposite fibers through controlled calcinations outlined in the experimental section. Figure 5 displays morphological and structural information of the typical MCNF-2 sample initially originating from MPF-2. Amazingly, the averaged diameter of MCNF-2 does not shrink apparently with respect to MPF-2 in spite of about 80 wt.% weight loss during the calcinations and this phenomenon is tentatively ascribed to formed carbon skeleton transformed from unprecedented heavily loaded polymer in FS recipe in contrast to electrospinning ingredients. Figure 5A&B show the micro/nanostructures of the nanocomposite fibers. The well-inherited intercrossing network of nanocomposites ensures great percolation of electrolyte and sufficient mechanical stability for volumetric expansion during battery-like pseudocapacitive TMOs performing. In the meanwhile, MCNFs exhibit undulate surface topography which is favorable to offer more accessible electrochemically active sites for Faradic redox reactions on the interface of electrolyte and electrode materials. The Figure 5C shows a cross-section image, taken under back-scatter detector, of an individual MCNF. Compared to lightweight carbon atoms, manganese with larger atomic number is chosen to be detected under backscattered electron imaging and apparently congregates on the surface or near surface regions of MCNFs,[32-33] unlike other homogeneous 1D hybrids.[16, 34] The formation of heterogeneous structures with TMONC enriched surface is essential for high rate pseudocapacitors because it is widely accepted that the pseudocapacitive reactions only effectively occur in the first dozens of nanometers at high current densities.[35-36] In other words, the utilization efficiency of pseudocapacitive materials in 1D configuration will decrease along the radical direction from exterior surface to interior axis. Heterogeneous 1D nanocomposites with active material-enriched surface or near surface utilize active materials most thoroughly. TEM studies of MCNF-2 were conducted to offer a further comprehension of structural features. In Figure D&E, the MCNFs display small, dark domains resulting from nanocrystals and relatively translucent carbon matrix, which indicates that the manganese oxide nanoparticles with about 10 nm size are evenly distributed and incorporated into near surface terrains of carbon frameworks and shielded by the surrounding meandering-featured carbon.[16] The meandering-featured carbon is believed to possess a graphitic-like structure that formed in relatively low temperature carbonization due to Mn^{2+} as oxidant catalyzing the carbonation of polymer.[37-39] The carbon protection can prevent oxide domains from further growth during material synthesis and undesired aggregation during long term cycles of charge and discharge, which is one of key structural characteristics of these unique nanocomposites to overcome the devastating capacitance fading in current designs of pseudocapacitors. The XRD pattern in Figure 5F indicates that manganosite-type MnO (JCPDS card #75–0626; space group: Fm3m (225); a=4.4435 Å) is readily identified by the emergence of five characteristic peaks (2θ = 34.94, 40.58, 58.72, 70.20 and 73.82°), marked by their corresponding Miller indices ((111), (200), (220), (311), and (222)).[40] This result is further confirmed by two diffraction rings, ascribed to (111), and (220) planes of MnO, in SAED pattern and the well-resolved lattice fringes with an interplanar spacing of 0.26 nm of the (111) plane of cubic MnO in HRTEM image (Figure 5E). The broad peak centered at about 25° is assigned to amorphous or graphitic-like carbon.[37-39]

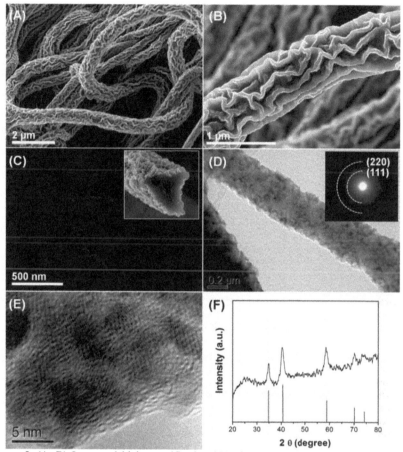

Figure 5. (A, B) Low- and high-magnification SEM images, (C) BackScatter image with inset of corresponding InLens image, (D, E) TEM and HRTEM images with inset of related SAED pattern in (E), and (F) XRD pattern of MnO nanocrystal-encased carbon nanocomposite fibers (MCNF-2).

The results of microscopy and XRD investigation are combined to propose a formation mechanism through which the heterogeneous MCNFs arise. The $Mn(NO_3)_2$/PVP composite MPFs become more and more viscous and start to plasticize with increasing temperature from room temperature to dehydration temperature of $Mn(NO_3)_2 \cdot xH_2O$. As the temperature continues increasing, the anhydrous $Mn(NO_3)_2$ starts to incline to diffuse outward due to the premature rigid shell of composite fibers causing different inward and outward ionic diffusion rates and subsequently decomposes into MnO_2 or other possible manganese compounds in argon with the simultaneous breaking down of the pendants on the PVP backbone.[41] After that, burst-nucleation of MnO_2 or other possible manganese compounds occurs once their concentration exceeds a critical concentration at certain temperature. And then, they continue to grow following the Ostwald ripening mechanism. During this process, however, the polymer matrix prevents them from further freely growing and aggregating. Finally, during the carbonization of PVP, MnO NCs are obtained through the reduction of manganese oxide with higher oxidation states by the surrounding abundant carbon. The collapse of the skin of carbon matrix and the phase separation of MnO NCs and carbon matrix are possible reasons for the formation of undulate surface morphology of these MCNFs.[42] To our best knowledge, it is the first time to form heterogeneous MnO NC-encased carbon nanocomposite fibers with the undulate surface morphology. In contrast, for electrospinning, in order to have relative low viscous and high dielectric spinning solution, it is not possible to have such high content of polymer, which is important to construct hierarchical nanocomposite carbon fibers.

In order to perform these heterogeneous structures as promising pseudocapacitive materials, in situ electrochemical oxidations are conducted in three-electrode cell system to phase transform electrochemically inactive MnO to pseudocapacitive MnO_2/MnO_x by applying one cycle of anodic charge on the MCNFs carrying electrodes at a current density of 10 μA cm^{-1} between -0.6 to 0.9 V for approximate 5000 s, and then 1000 cycles of voltammetric scans with a potential window of 0 – 1 V (vs Ag/AgCl). According to Messaoudi et al.,[43] the transition sequence is described by the following stoichiometric reaction equations with slight modifications, even though no particular transition would happen exclusively and thoroughly in each stage.

$$3MnO + H_2O \leftrightarrow Mn_3O_4 + 2H^+ + 2e^-$$
$$Mn_3O_4 \cdot 2H_2O + OH^- \leftrightarrow 2MnOOH + Mn(OH)_3 + e^-$$
$$4MnOOH + 2Mn(OH)_3 + 3OH^- \leftrightarrow 6MnO_2 + 5H_2O + 3H^+ + 6e^-$$

Figure 6A exhibits the evolutions of CV curves and remarkable uphill leap in higher potential regions and expanded CV integrated areas are observed during the first three cyclic voltammetric oxidations, which is an evident indication of electrochemical oxidations of MnO and increased specific capacitances.[44-45] Figure 6B compares the CV curves of MCNF-2 before electrochemical oxidations and after anodic oxidation and 1000 cycles of CV oxidations, respectively. Before electrochemical oxidations, the specific capacitance (C_s) of MCNF-2 is estimated to be 3.4 F g^{-1} from the calculation of integrated area of CV curve. Partial contribution of the C_s can be attributed to the double-layer energy storage that arises from electrostatic adsorption of electrolyte ions when hierarchical carbon skeleton with enhanced surface area is taken into account. Particularly noteworthy is that after anodic oxidation at an areal current density of 10 μm cm^{-1} and oxidations of 1000 CV potential cycles, the C_s is dramatically boosted to 100.4 and 128.3 F g^{-1}, respectively. This can be explained by the phase transformation from inactive MnO to pseudocapacitive MnO_2/MnO_x which activates pseudocapacitive energy storage on the surface or near surface of nanocomposites and eventually converts these unique heterogeneous nanostructures into pseudocapacitive electrode materials. Galvanostatic charge and discharge measurements are carried out to further investigate nanocomposites at a variety of current densities. The excellent symmetry of charge and discharge times and the pyramidal waves indicate a superior reversibility of the nanocomposites. The potential plateaus at about 0.7 V and 0.4 V on pyramids are well consistent with the anodic and cathodic peaks in CV studies (Figure 6B) and validate the Faradic redox reactions.[46-47] Figure 6D shows the C_s, estimated from discharging times, as a function of the different discharging current densities. The nanocomposites deliver a maximum C_s of 257.9 F g^{-1} at a current density of 0.2 A g^{-1}. Also, the heterogeneous nanocomposite fibers maintain a C_s of 214.2 F g^{-1} at a high current density of 1.0 A g^{-1}, which is about 83.1 % of the value at 0.2 A g^{-1}. The fiber-twined network with kinetically unlimited ion diffusion is mainly responsible for the extraordinary rate capacity. The slope of linear correlation between IR drops and discharging current density can be applied to evaluate the internal resistance of electrode materials.[48] It is interesting to notice that the nanocomposites exhibit a low internal resistance without employing conductive and polymeric additives, which is owing to the firm attachment between nanocomposite electrode materials and current collector. Making use of high speed centrifugal force and wettability of PVP plays an important role in fabrication of mechanically stable electrodes.

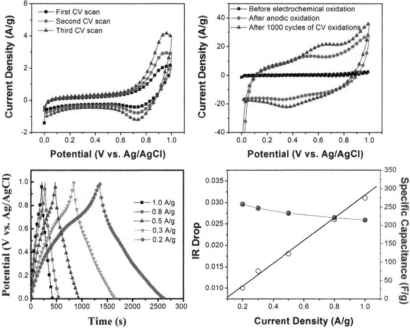

Figure 6. (A) The first three cyclic voltammetric cycles (CV) of MCNF-2 at a scan rate of 100 mV s^{-1}. (B) CV evolutions before electrochemical oxidations and after anodic and 1000 cycles of CV oxidations at a scan rate of 50 mV s^{-1}. (C) Galvanostatic charge and discharge curves at various current densities, and (D) specific capacitances and IR drops as a function of the current densities of electrochemically oxidized MCNF-2 (EO-MCNF-2).

CONCLUSION

In summary, we have successfully fabricated nanoarchitectured heterogeneous MnO$_x$ NC-encased carbon nanocomposite fibers with MnO$_x$ NC-enriched surface via a novel FS method followed by calcination and electrochemical oxidations. The controllable ionic diffusion during thermolysis and phase transformation from electrochemically inactive MnO NCs to pseudocapacitive MnO_x NCs by anodic and CV scan oxidations has near surface regions of nanocomposite fibers incorporated with MnO_x NCs accessible to Faradic redox reactions that occur on the surface/near surface. The nanocomposite electrode with electrochemically oxidative enhancements can achieve boosted C_s of 128.3 F g^{-1} from initial value of 3.4 F g^{-1} before oxidations at a scan rate of 100 mV s^{-1}. The Mn^{2+} cations as oxidizing agent can accelerate carbonation of polymeric precursor and make carbonation take place at relatively low temperature condition. The fiber-twined network enables a high rate capacity in view of ideal pathways for ion percolation and electron transportation without kinetic limitations. When current density is up to 1.0 A g^{-1}, the C_s still maintain about 83.1 % of the maximum C_s obtained at 0.2 A g^{-1}. Owing to functional oxide nanoparticles thoroughly shielded by thin layer conductive carbon, the unique heterogeneous nanocomposite fibers are expected to resolve the uncontrollable aggregation and pulverization problems of oxide nanoparticles and become one of the best electrode designs for high performance pseudocapacitors.

ACKNOWLEDGEMENT
    The authors appreciate the support from the University of Texas-Pan American (startup for YM), American Chemical Society-Petroleum Research Fund #51497 (YM), and the National Science Foundation under DMR grant # 0934157 (PREM-UTPA/UMN-Science and Engineering of Polymeric and Nanoparticle-based Materials for Electronic and Structural Applications) and DMR MRI grant #1040419.

REFERENCES
[1]J. S. Lee, O. S. Kwon, S. J. Park, E. Y. Park, S. A. You, H. Yoon, and J. Jang, Fabrication of Ultrafine Metal-Oxide-Decorated Carbon Nanofibers for DMMP Sensor Application, *ACS Nano,* **5**, 7992-8001 (2011).
[2]I. Capek, Dispersions, Novel Nanomaterial Sensors and Nanoconjugates Based on Carbon Nanotubes, *Adv Colloid Interfac,* **150**, 63-89 (2009).
[3]Y. Liang, Y. Li, H. Wang, and H. Dai, Strongly Coupled Inorganic/Nanocarbon Hybrid Materials for Advanced Electrocatalysis, *J Am Chem Soc,* **135**, 2013-36 (2013).
[4]A. Jensen, J. R. Hauptmann, J. Nygard, J. Sadowski, and P. E. Lindelof, Hybrid Devices from Single Wall Carbon Nanotubes Epitaxially Grown into a Semiconductor Heterostructure, *Nano Lett,* **4**, 349-352 (2004).
[5]R. Liu, J. Duay, and S. B. Lee, Redox Exchange Induced MnO_2 Nanoparticle Enrichment in Poly(3,4-ethylenedioxythiophene) Nanowires for Electrochemical Energy Storage, *ACS Nano,* **4**, 4299-4307 (2010).
[6]X. Xiao, T. Li, P. Yang, Y. Gao, H. Jin, W. Ni, W. Zhan, X. Zhang, Y. Cao, J. Zhong, L. Gong, W. C. Yen, W. Mai, J. Chen, K. Huo, Y. L. Chueh, Z. L. Wang, and J. Zhou, Fiber-Based All-Solid-State Flexible Supercapacitors for Self-powered Systems, *ACS Nano,* **6**, 9200-6 (2012).
[7]J. Duay, E. Gillette, R. Liu, and S. B. Lee, Highly Flexible Pseudocapacitor Based on Freestanding Heterogeneous MnO_2/Conductive Polymer Nanowire Arrays, *Phys Chem Chem Phys,* **14**, 3329-3337 (2012).
[8]R. H. Baughman, A. A. Zakhidov, and W. A. de Heer, Carbon Nanotubes - the Route toward Applications, *Science,* **297**, 787-792 (2002).
[9]Y. Y. Liang, M. G. Schwab, L. J. Zhi, E. Mugnaioli, U. Kolb, X. L. Feng, and K. Mullen, Direct Access to Metal or Metal Oxide Nanocrystals Integrated with One-Dimensional Nanoporous

Carbons for Electrochemical Energy Storage, *Journal of the American Chemical Society,* **132,** 15030-15037 (2010).

[10]S. B. Ma, K. W. Nam, W. S. Yoon, X. Q. Yang, K. Y. Ahn, K. H. Oh, and K. B. Kim, Electrochemical Properties of Manganese Oxide Coated onto Carbon Nanotubes for Energy-Storage Applications, *J Power Sources,* **178,** 483-489 (2008).

[11]S. Boukhalfa, K. Evanoff, and G. Yushin, Atomic Layer Deposition of Vanadium Oxide on Carbon Nanotubes for High-power Supercapacitor Electrodes, *Energ Environ Sci,* **5,** 6872-6879 (2012).

[12]M. Sathiya, A. S. Prakash, K. Ramesha, J. M. Tarascon, and A. K. Shukla, V_2O_5-Anchored Carbon Nanotubes for Enhanced Electrochemical Energy Storage, *Journal of the American Chemical Society,* **133,** 16291-16299 (2011).

[13]W. Tang, X. W. Gao, Y. S. Zhu, Y. B. Yue, Y. Shi, Y. P. Wu, and K. Zhu, A Hybrid of V_2O_5 Nanowires and MWCNTs Coated with Polypyrrole as an Anode Material for Aqueous Rechargeable Lithium Batteries with Excellent Cycling Performance, *J Mater Chem,* **22,** 20143-20145 (2012).

[14]L. W. Ji, Z. Lin, M. Alcoutlabi, O. Toprakci, Y. F. Yao, G. J. Xu, S. L. Li, and X. W. Zhang, Electrospun Carbon Nanofibers Decorated with Various Amounts of Electrochemically-Inert Nickel Nanoparticles for Use as High-Performance Energy Storage Materials, *Rsc Adv,* **2,** 192-198 (2012).

[15]S. R. Li, Y. Sun, S. Y. Ge, Y. Qiao, Y. M. Chen, I. Liebervvirth, Y. Yu, and C. H. Chen, A Facile Route to Synthesize Nano-MnO/C Composites and Their Application in Lithium Ion Batteries, *Chem Eng J,* **192,** 226-231 (2012).

[16]L. W. Ji, A. J. Medford, and X. W. Zhang, Porous Carbon Nanofibers Loaded with Manganese Oxide Particles: Formation Mechanism and Electrochemical Performance as Energy-Storage Materials, *J Mater Chem,* **19,** 5593-5601 (2009).

[17]K. Sarkar, C. Gomez, S. Zambrano, M. Ramirez, E. de Hoyos, H. Vasquez, and K. Lozano, Electrospinning to Forcespinning (TM), *Mater Today,* **13,** 12-14 (2010).

[18]K. Shanmuganathan, Y. C. Fang, D. Y. Chou, S. Sparks, J. Hibbert, and C. J. Ellison, Solventless High Throughput Manufacturing of Poly(butylene terephthalate) Nanofibers, *ACS Macro Lett,* **1,** 960-964 (2012).

[19]Q. P. Pham, U. Sharma, and A. G. Mikos, Electrospinning of Polymeric Nanofibers for Tissue Engineering Applications: A Review, *Tissue Eng,* **12,** 1197-1211 (2006).

[20]J. M. Du, S. Shintay, and X. W. Zhang, Diameter Control of Electrospun Polyacrylonitrile/Iron Acetylacetonate Ultrafine Nanofibers, *J Polym Sci Pol Phys,* **46,** 1611-1618 (2008).

[21]O. Dag, O. Samarskaya, C. Tura, A. Gunay, and O. Celik, Spectroscopic Investigation of Nitrate-Metal and Metal-Surfactant Interactions in the Solid $AgNO_3/C_{12}EO_{10}$ and Liquid-Crystalline $[M(H_2O)_n](NO_3)_2/C_{12}EO_{10}$ Systems, *Langmuir,* **19,** 3671-3676 (2003).

[22]J. Y. Xian, Q. Hua, Z. Q. Jiang, Y. S. Ma, and W. X. Huang, Size-Dependent Interaction of the Poly(N-vinyl-2-pyrrolidone) Capping Ligand with Pd Nanocrystals, *Langmuir,* **28,** 6736-6741 (2012).

[23]M. A. Moharram, and M. G. Khafagi, Application of FTIR Spectroscopy for Structural Characterization of Ternary Poly(acrylic acid)-Metal-Poly(vinyl pyrrolidone) Complexes, *J Appl Polym Sci,* **105,** 1888-1893 (2007).

[24]L. N. C. Rodrigues, M. G. Issa, A. C. C. Asbahr, M. A. C. Silva, and H. G. Ferraz, Multicomponent Complex Formation between Pyrimethamine, Cyclodextrins and Water-Soluble Polymers, *Braz Arch Biol Techn,* **54,** 965-972 (2011).

[25]G. Gaucher, K. Asahina, J. H. Wang, and J. C. Leroux, Effect of Poly(N-vinyl-pyrrolidone)-Block-Poly(D,L-lactide) as Coating Agent on the Opsonization, Phagocytosis, and Pharmacokinetics of Biodegradable Nanoparticles, *Biomacromolecules,* **10**, 408-416 (2009).
[26]C. B. Jing, X. G. Xu, X. L. Zhang, Z. B. Liu, and J. H. Chu, In Situ Synthesis and Third-Order Nonlinear Optical Properties of CdS/PVP Nanocomposite Films, *J Phys D Appl Phys,* **42**, (2009).
[27]C. K. Chan, I. M. Chu, C. F. Ou, and Y. W. Lin, Interfacial Interactions and their Influence to Phase Behavior in Poly(vinyl pyrrolidone)/Silica Hybrid Materials Prepared by Sol-gel Process, *Mater Lett,* **58**, 2243-2247 (2004).
[28]X. Chen, K. M. Unruh, C. Y. Ni, B. Ali, Z. C. Sun, Q. Lu, J. Deitzel, and J. Q. Xiao, Fabrication, Formation Mechanism, and Magnetic Properties of Metal Oxide Nanotubes via Electrospinning and Thermal Treatment, *J Phys Chem C,* **115**, 373-378 (2011).
[29]J. M. Kim, H. I. Joh, S. M. Jo, D. J. Ahn, H. Y. Ha, S. A. Hong, and S. K. Kim, Preparation and Characterization of Pt Nanowire by Electrospinning Method for Methanol Oxidation, *Electrochim Acta,* **55**, 4827-4835 (2010).
[30]K. X. Yao, and H. C. Zeng, ZnO/PVP Nanocomposite Spheres with Two Hemispheres, *J Phys Chem C,* **111**, 13301-13308 (2007).
[31]L. C. Mendes, R. C. Rodrigues, and E. P. Silva, Thermal, Structural and Morphological Assessment of PVP/HA composites, *J Therm Anal Calorim,* **101**, 899-905 (2010).
[32]M. Mohl, D. Dobo, A. Kukovecz, Z. Konya, K. Kordas, J. Q. Wei, R. Vajtai, and P. M. Ajayan, Formation of CuPd and CuPt Bimetallic Nanotubes by Galvanic Replacement Reaction, *J Phys Chem C,* **115**, 9403-9409 (2011).
[33]F. Xia, J. Brugger, Y. Ngothai, B. O'Neill, G. R. Chen, and A. Pring, Three-Dimensional Ordered Arrays of Zeolite Nanocrystals with Uniform Size and Orientation by a Pseudomorphic Coupled Dissolution-Reprecipitation Replacement Route, *Cryst Growth Des,* **9**, 4902-4906 (2009).
[34]L. W. Ji, O. Toprakci, M. Alcoutlabi, Y. F. Yao, Y. Li, S. Zhang, B. K. Guo, Z. Lin, and X. W. Zhang, alpha-Fe_2O_3 Nanoparticle-Loaded Carbon Nanofibers as Stable and High-Capacity Anodes for Rechargeable Lithium-Ion Batteries, *ACS Appl Mater Inter,* **4**, 2672-2679 (2012).
[35]P. Simon, and Y. Gogotsi, Materials for Electrochemical Capacitors, *Nature Materials,* **7**, 845-854 (2008).
[36]H. Jiang, C. Z. Li, T. Sun, and J. Ma, A Green and High Energy Density Asymmetric Supercapacitor based on Ultrathin MnO_2 Nanostructures and Functional Mesoporous Carbon Nanotube Electrodes, *Nanoscale,* **4**, 807-812 (2012).
[37]Y. M. Sun, X. L. Hu, W. Luo, and Y. H. Huang, Porous Carbon-Modified MnO Disks Prepared by a Microwave-Polyol Process and their Superior Lithium-ion Storage Properties, *J Mater Chem,* **22**, 19190-19195 (2012).
[38]J. X. Zhu, T. Sun, J. S. Chen, W. H. Shi, X. J. Zhang, X. W. Lou, S. Mhaisalkar, H. H. Hng, F. Boey, J. Ma, and Q. Y. Yan, Controlled Synthesis of Sb Nanostructures and Their Conversion to $CoSb_3$ Nanoparticle Chains for Li-Ion Battery Electrodes, *Chemistry of Materials,* **22**, 5333-5339 (2010).
[39]Z. Y. Wang, J. S. Chen, T. Zhu, S. Madhavi, and X. W. Lou, One-pot Synthesis of Uniform Carbon-coated MoO_2 Nanospheres for High-rate Reversible Lithium Storage, *Chem Commun,* **46**, 6906-6908 (2010).

[40]X. W. Li, D. Li, L. Qiao, X. H. Wang, X. L. Sun, P. Wang, and D. Y. He, Interconnected Porous MnO Nanoflakes for High-Performance Lithium Ion Battery Anodes, *J Mater Chem,* **22**, 9189-9194 (2012).

[41]M. Maneva, and N. Petroff, The Thermal Dehydration, Decomposition and Kinetics of $Mn(NO_3)_2.6H_2O$ and Its Deuterated Analog, *J Therm Anal,* **36**, 2511-2520 (1990).

[42]Y. Ding, Y. Wang, L. C. Zhang, H. Zhang, and Y. Lei, Preparation, Characterization and Application of Novel Conductive NiO-CdO Nanofibers with Dislocation Feature, *J Mater Chem,* **22**, 980-986 (2012).

[43]B. Messaoudi, S. Joiret, M. Keddam, and H. Takenouti, Anodic Behaviour of Manganese in Alkaline Medium, *Electrochim Acta,* **46**, 2487-2498 (2001).

[44]B. Djurfors, J. N. Broughton, M. J. Brett, and D. G. Ivey, Electrochemical Oxidation of Mn/MnO Films: Mechanism of Porous Film Growth, *J Electrochem Soc,* **153**, A64-A68 (2006).

[45]H. Xia, Y. S. Meng, X. G. Li, G. L. Yuan, and C. Cui, Porous Manganese Oxide Generated from Lithiation/Delithiation with Improved Electrochemical Oxidation for Supercapacitors, *J Mater Chem,* **21**, 15521-15526 (2011).

[46]Z. B. Lei, F. H. Shi, and L. Lu, Incorporation of MnO_2-Coated Carbon Nanotubes between Graphene Sheets as Supercapacitor Electrode, *ACS Appl Mater Inter,* **4**, 1058-1064 (2012).

[47]X. Y. Lang, A. Hirata, T. Fujita, and M. W. Chen, Nanoporous Metal/Oxide Hybrid Electrodes for Electrochemical Supercapacitors, *Nat Nanotechnol,* **6**, 232-236 (2011).

[48]C. W. Huang, and H. S. Teng, Influence of Carbon Nanotube Grafting on the Impedance Behavior of Activated Carbon Capacitors, *J Electrochem Soc,* **155**, A739-A744 (2008).

THE EFFECT OF GEOMETRIC FACTORS ON SODIUM CONDUCTION: A COMPARISON OF BETA- AND BETA"-ALUMINA

Emma Kennedy and Dunbar P. Birnie III
Materials Science and Engineering, Rutgers University
Piscataway, NJ, USA

ABSTRACT

β"-alumina has been shown to have superior sodium conduction compared with β-alumina, and therefore is used as the electrolyte material in sodium battery systems. The β"-alumina structure is composed of layers of (111) oriented spinel-structure units separated by conduction planes that are comparably more open allowing for sodium diffusion. The β-alumina structure is similar except that the spinel units are aligned differently creating a smaller and less flexible conduction plane. In addition to being more flexible, β"-alumina is able to absorb stabilizing ions into the spinel unit, whereas β-alumina does not appear to have that capability. In this paper we analyze the differences in these structures that affect sodium conduction. This comparison will assist in understanding the β"-alumina structure and enhancing its capacity for greater conduction of sodium.

INTRODUCTION

Today β"-alumina is the preferred electrolyte material for sodium battery systems compared to β-alumina. This preference is a result of the ability of sodium to migrate through the conduction plane of the electrolyte, the β" conduction plane is more flexible and open. The differences between these two structures can be seen in Figure 1, the structure for the β"-alumina is a four layer ABCA stack that is spinel-like in its structure and a conduction slab that has fewer oxygen atoms and more room for sodium ion motion[1,2]. The β-alumina structure is similar but there is a 180° rotation from the first set of four oxygen layers to the second set of oxygen layers.[3,4] The rotation causes the main issue with the flow of sodium atoms in the conduction plane because at a few points of the conduction pathway there are two oxygen three (O(3)) atoms that sit on top and below the diffusing atom, this point is known as the anti Beevers-Ross site. The β" structure does not have this problem because instead of having an O(3) directly on top of another O(3) atom the O(3) is on top or below the center of a cluster of three O(4) atoms.[5] Another difference that the two structures have is that the diffusion pathway in the β" structure changes in the z direction and in the β the diffusing atoms stay on the same z plane, and this could also be a factor in the flexibility of the structure.

The pressure from surrounding oxygen atoms was examined as the sodium atoms moves from one spot to another in the hexagonal pattern that both structures have, by doing this we were able to see the points that affect the conduction the most in both structures. We also calculated the height of the conduction plane and the oxygen block to see how that was affected by the diffusion of sodium.

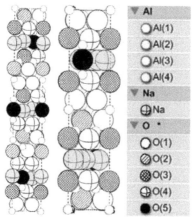

Figure 1. A comparison of β"-alumina (left) and β-alumina (right), on the far right is the key for both images and the image in Figure 2.

EXPERIMENTAL

The β"-alumina structure was reported by Bettman and Peters and has a space group of R-3m, and a hexagonal cell with a_o = 5.614 Å and c_o = 33.85 Å.[1] The atom locations for β" can be found in Table I. The β-alumina structure is from Peters and Bettman as well, the space group for this structure is 6_3/mmc, it is also hexagonal with a_o = 5.594Å and c_o = 22.53Å.[3,5] The atom locations for this structure are in Table II. There are three major sites in the β-alumina conduction pathway, the first is the anti Beevers-Ross (aBR) site, this site is the very center on both images in Figure 2, the second site is the Beevers-Ross (BR) site and it has four location on Figure 2a, it is the point at the center of the tight clusters of three sodium atoms, the third site is the mid oxygen point and that is most obviously seen on Figure 2b where the 2 is labeled on the image.[3,6] The location for the sodium atoms in Figure 2a came from the electron density map in Figure 2b, instead of being on one of the major locations the sodium atoms are located where they are most likely to be found, surrounding the BR site and close to the mid oxygen point, and not on the aBR site.

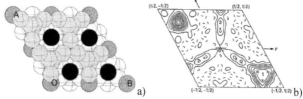

Figure 2. a) The conduction plane for sodium in the β-alumina structure (the key for Figure 1 applies to this image as well). b) The Fourier section showing the electron density on the conduction plane of β-alumina[3]. The 1 represents the Beevers-Ross site, the 2 represents the mid-oxygen point, the X in the middle represents the anti-Beevers-Ross site, and the x's are where the highest electron density is and where the Na ions are located in a.

Table I. Atom locations for the β"-alumina structure[1]

Atom	# in Cell and Wyckoff Position	X	Y	Z
Na	6c	0	0	0.1717
Al(1)	3a	0	0	0
Al(2)	6a	0	0	0.3501
Al(3)	18h	0.3362	0.1681	0.0708
Al(4)	6c	0	0	0.4498
O(1)	18h	0.1562	0.3124	0.0339
O(2)	6c	0	0	0.2955
O(3)	6c	0	0	0.0961
O(4)	18h	0.1657	0.3314	0.2357
O(5)	3b	0	0	0.5

Table II. Atom location for the β-alumina structure[3]

Atom	# in Cell and Wyckoff Position	X	Y	Z
Na(1)	12k	0.1571	-0.1571	0.0501
Na(2)	12k	0.5032	-0.5032	0.1468
O(1)	4f	0.6667	-0.6667	0.0555
O(2)	2c	0.3333	-0.3333	0.25
O(3)	12k	-0.1678	0.1678	0.1063
O(4)	4e	0	0	0.1425
O(5)	4f	0.3333	-0.3333	0.0248
Al(1)	4f	0.3333	-0.3333	0.1756
Al(2)	2a	0	0	0
Al(3)	6h	-0.2398	0.2398	0.25
Al(4)	6h	-0.1269	0.1269	0.25

In order to compare the structures ability to conduct sodium we used a program to simulate the two structures in order to look at the path the sodium would take diffusing through the electrolyte and its distances to the neighboring oxygen atoms as it diffuses. The images that resulted from this program are in Figures 1 and 2. In this program we were able to add sodium atoms in the hexagonal conduction pathway and then measure the distances between them and the O(5), O(4), and O(3) atoms. These distances were then graphed and the relationships that affected conduction the most for the β-alumina and the β"-alumina were compared in Figure 3.
In this figure, graph a is showing the bond distances of the sodium atom and different O(3) atoms in β"-alumina as it moves from one position to another and then to another, the diffusing ion starts close to one O(3) and as it moves away from that one it is getting closer to different O(3) atom and then it finally moves closer to a third O(3) atom. We only showed this relationship for the β" because this one has the shortest distance of all of oxygen atoms surrounding the conduction plane. Figure 2b is different because there is only one O(3) line and that is because the O(3) on the top and the O(3) on the bottom will always have the same bond distances, so

instead we look at the O(3) and the O(4). For this plot the starting position is at the BR site the middle position is at the aBR site and the final position is again at a different BR site.

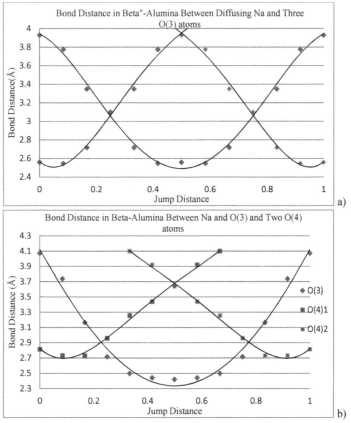

Figure 3. a) The bond distances between the diffusing ion and the O(3) as the ion moves to two different spots. The three lines represent the three O(3) atoms. b) The three lines show the bond distances of the O(3) atoms and the two O(4) atoms with the sodium as it starts at the BR site and moves to the aBR site and then to the next BR site.

Looking at these graphs we were able to see the points in the diffusion that put the most pressure on the diffusing atom and the point that is the tightest for both structures is found around the O(3) atom, the distance between the diffusing atom and the O(3) in the β" structure is about 2.53Å, and the distance for the β structures diffusing atom is 2.36Å. The difference between these two numbers is a result of the ability of the sodium in the β" structure to move in the z direction away from the O(3) atom, the sodium in the β structure does not have the space to move in this direction.

To look at the flexibility and the effects of charge in the structure we used Harbach's equations[4] to calculate the H and D length for both structures, these lengths can be seen in Figure 4. The H length is for the conduction plane and is done for the two oxygen atoms that surround the conduction plane, O(3) and O(4) atoms and the equations for these distances are:

$$H_3 = c_o * (1/3 - 2 * z(O3)) \tag{1}$$

$$H_4 = c_o * (2 * z(O4) - 1/3) \tag{2}$$

Where $z(O3)$ and $z(O4)$ are the z axis of one of the O(3) and O(4) atoms in the β"-alumina structure found on Table I. The equations for the β-alumina heights are very similar the only difference is instead of dividing the c_o by 3 it is divided by 2 because where the β" structure is made up of 3 blocks the β structure is only made up of 2. These equations are below.

$$H_3 = c_o * (1/2 - 2 * z(O3)) \tag{3}$$

$$H_4 = c_o * (2 * z(O4) - 1/2) \tag{4}$$

Next we used these H equations to find the D_{actual} value, using the equation below which we derived. The first one is for β"-alumina the second one is for β-alumina, and the third is the ideal D and is the same equation for both structures.

$$D_{actual} = {c_o}/3 - \frac{H_3 + 2H_4}{3} \tag{5}$$

$$D_{actual} = {c_o}/2 - \frac{H_3 + 2H_4}{3} \tag{6}$$

$$D_{ideal} = \frac{\sqrt{3}}{\sqrt{2}} * a_o \tag{7}$$

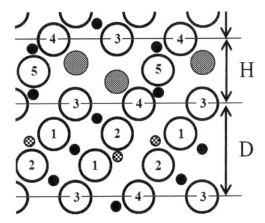

Figure 4. Crystal structure of β" alumina viewed from the side emphasizing the stacking arrangement of close-packed oxygen layers that make up the structure. Spinel-related blocks are identified with "D" and conduction slabs are identified by "H"[4].

RESULTS AND DISCUSSION

By analyzing the β and β"-alumina structures we were able to understand more clearly why the β" is the more preferred structure in hopes to one day improve on the sodium conduction. When looking at the physical structure we could see how the β structure has a position in the conduction pathway that has two O(3) atoms on top of one another. We believe this is what causes the main problem for conduction through this structure. In the β"-alumina structure there is a position with one O(3) atom on top of the conducting ion but because there is no atom directly below it the ion moves slightly down closer to the O(4) atoms to avoid the pressure from the O(3) atom and then moves back up to move around the next O(3) atom. A sodium ion in the β structure doesn't have this ability to move in the z direction because of the sandwiching effect of the O(3)s and the two sets of three O(4)s that are directly above and below the conduction pathway. This means the ion is on the same plane through the entire conduction path and there is no flexibility of space in the z direction.

Using the Harbach Equations above we were able to calculate the H and D values for both structures and those calculations can be seen in Table III. What we expected to find was the the D_{actual} for the β"-alumina would be smaller than the β-alumina because of electrostatic effects caused by the charge of the sodium ions in the conduction plane and the charge of the stabilizing atoms in the oxygen block which would result in a squeezing effect on the block. This was not the case, the β structure might be affected by oxygen vacancies in the conduction plane or impurities in the bulk. The H_3 for the β-alumina was larger than the β"-alumina because the β doesn't have much room to close that gap because of the O(3) that are on top of each other whereas the O(3) in the β" structure doesn't have that restriction. The major difference in these values between these two structures is the difference between the H_3 and the H_4 in the β-alumina and the H_3 and H_4 in the β"-alumina, the β has a larger difference and that again is due to this inability to compress the stack of atoms on top of one another. The relationship between the D_{actual} and the D_{ideal} were what we expected, the D_{actual} came out to be around 4.4% less than the ideal and that is because of the electrostatic effects in the structure.

Table III. Results of calculation of Harbach equations for β- and β"-alumina

	β-Alumina	β"-Alumina
a_0	5.594	5.614
c_0	22.53	33.85
z(O3)	0.1425	0.0961
z(O4)	0.1467	0.2357
H_3	4.8439	4.7773
H_4	4.6510	4.6735
D_{actual}	6.5496	6.5751
D_{ideal}	6.8512	6.8757

CONCLUSION

A comparison of β- and β"-alumina was done to get a better understanding of the material that is regularly used as the electrolyte material in sodium battery systems. Using a simulated structure we were able to look at the atoms surrounding the conduction plane and measure the pressure that they might apply to the conducting ion. The biggest difference in the two structures was that at certain points along the pathway the ion would have to squeeze through two overlapping O(3) atoms, and then the rest of the pathway is clear of obstruction. The β" has pretty much equal pressure all along the pathway, the O(3) atoms in this structure are also the tightest spot but in this structure the ion is able to move in the z direction to avoid a tight squeeze because there is no atom directly below it. We also examined the height of the conduction plane and the oxygen block to get an idea of how those were affected by the charges in the structure as well as the two different orientations of the structures. What we found with those calculations was not what we expected, and we have to assume that the β structure is more complicated than we thought. Our goal for this research is to better understand the β" structure so that someday we can improve its conduction and build a better battery.

ACKNOWLEDGMENTS
The authors are grateful for support from the HED Graduate Fellowship and the McLaren Chair Endowment.

REFERENCES
[1] M. Bettman, C.R. Peters, The Crystal Structure of $Na_2O \cdot MgO \cdot 5Al_2O_3$ with Reference to $Na_2O \cdot 5Al_2O_3$ and Other Isotypal Compounds, *The Journal of Physical Chemistry*, 73, 1774-1780, (1969)

[2] D.P. Birnie III, On the Structural Integrity of the Spinel Block in the Beta"-alumina structure, *Acta Crystallographica*, B68, 118-122, (2012)

[3] C.R. Peters, M. Bettman, J.W. Moore, M.D. Glick, Refinement of the Structure of Sodium Beta-Alumina, *Acta Crystallographica*, B27, 1826, (1971)

[4] W.L. Bragg, C. Gottfried, J. West, The Structure of Beta Alumina, *Zeitschrift für Kristallographie*, 77, 255-274, (1931)

[5] X. Lu, G. Xia, J.P. Lemmon, Z. Yang, Advanced Materials for Sodium-Beta Alumina Batteries: Status, Challenges, Perspectives, *Journal of Power Sources*, 195, 2431-2442, (2010)

[6] H.C. Brinkhoff, Substitution of Aluminum by Gallium in Beta-Alumina, *The Journal of Physical Chemistry*, 35, 1225-1229, (1974)

Advanced Ceramic Materials and Processing for Photonics and Energy

EFFECT OF POROSITY ON THE EFFICIENCY OF DSSC PRODUCED
BY USING NANO-SIZE TiO$_2$ POWDERS

N. Bilgin[1*], J. Park[2], A. Ozturk[1]

[1]Middle East Technical University, Metallurgical and Materials Engineering Department, Ankara
06800, Turkey.

[2]Atilim University, Metallurgical and Materials Engineering Department, Ankara 06836, Turkey.

ABSTRACT

The effect of porosity on the energy conversion efficiency of dye-sensitized solar cells
(DSSCs) prepared by using pastes formed by mixing 20 nm and 200 nm TiO$_2$ particles in different
ratios is investigated. XRD and SEM analysis have been done to investigate the microstructure of
pastes. The energy conversion efficiency of DSSCs was determined by drawing complete current
density-voltage curve. The DSSC prepared using the mixture composed of 40 wt% 20 nm and 60
wt% 200 nm TiO$_2$ particles maintained best energy conversion efficiency of 6.74%. The energy
conversion efficiency of the DSSCs prepared by using pastes based on the mixture of two different
size of TiO$_2$ particles is much better than that of the DSSCs prepared by using pastes composed of
either only 20 nm or only 200 nm of TiO$_2$ particles. The improved energy conversion efficiency is
attributed to the establishment of further porous structure that lets more dye absorption from the
surface through interior which provides enhancement of light absorption and multiple scattering.

Keywords: Porosity, Dye-sensitized solar cell, TiO$_2$, Conversion efficiency

INTRODUCTION

Recently, the dye-sensitized solar cell (DSSC) has received considerable attention as a
promising energy producer because of low production cost and low environmental damage during
fabrication.[1-5] TiO$_2$ particles have been commonly used as the main component of the
photoelectrode system in most DSSCs. The efficiency of the DSSC depends mainly on two factors:
the absorption spectrum of the dye and the anchorage of the dye to the surface of TiO$_2$.[6-8] The
particle size of TiO$_2$ powder influences the performance of the solar cell. The most efficient
electrodes in DSSCs include of a porous TiO$_2$ layer approximately 10-μm thick with an
interconnected network of nano-size crystals roughly 20 nm in size.[9,10] This layer has a smaller
surface area with respect to larger particle sized films, possibly reducing the amount of light

absorbed and the number of electrons and holes generated. Electrodes with larger particles have larger contact points which means in larger particles included films, each particle touch themselves much more points, between sintered particles, allowing for easier dye access and better dye assembly. Also layers with larger particles contain bigger porosity which decreases the number of possible percolation paths. Conversely, the electrode with smaller particles have larger surface area and more contact points between sintered particles, allowing for better dye adsorption by this way there are less dead ends. In contrast, this layer exhibits a larger number of grain boundaries that electrons need to pass through, which results in a higher probability of electron trapping.[11-14]

Although many investigations have been conducted on DSSC, the effect of porosity resulted by TiO$_2$ layer on conversion efficiency of DSSC still remains mysterious. This study has been devoted to prepare photoelectrodes composed of two different sizes of TiO$_2$ particles, and to determine the effect of porosity on energy conversion efficiency. The performance of DSSC was measured by I-V test. The microstructural morphology of the electrodes was examined by scanning electron microscope and atomic force microscopy.

EXPERIMENTAL

Preparation of Working and Counter Electrode

Two different size, small (20 nm) and large (200 nm), commercially available TiO$_2$ particles were used for the preparation of pastes. Small TiO$_2$ particles were P25 powder (anatase and rutile) supplied from Degussa whereas large TiO$_2$ particles were supplied from Alfa Aesar. Photoelectrode pastes were formed first by mixing both powders at different ratios. Total weight of each mixture was 3 g. Then, each mixture was blended with 0.05 g 4-hydroxy benzoic acid and 4 mL ethanol under sonication for 5 min. After that, the blend was dried at 70°C for 30 min. After drying, 0.03 g ethyl cellulose and 5 mL α-terpineol were added homogeneously to form a paste. Code of photoelectrode pastes and their small and large size TiO$_2$ particle weight percentage (wt%) are tabulated in Table 1.

Table 1. Photoelectrode pastes and their percentages

Paste Code	20 nm TiO$_2$ (wt %)	200 nm TiO$_2$ (wt %)
A	100	0
B	80	20
C	60	40
D	40	60
E	20	80
F	0	100

A layer of the photoelectrode paste was coated onto the fluorine-doped tin oxide (FTO) glass (15Ω/square) by screen printing method. Then, the FTO glass was dried in an oven at 130 °C for 10 min. This procedure was repeated 3 times for each paste to obtain an electrode. The electrode prepared was sintered at 500°C for 30 min to establish an interconnection between the TiO$_2$ particles and to remove any remaining organic materials. Final thickness of the samples are between 15-25 µm. The area of the working electrode was 0.25 cm^2. The electrode was immersed in a 3 mM dye solution (N719, Solaronix) at room temperature for 12 h. Finally, it was rinsed with ethanol to remove the excess dye on the titania layer. In order to prepare the counter electrode, first two holes were drilled in the FTO glass. Then a Pt solution was prepared by mixing H$_2$PtCl$_6$ powder with IPA liquid. Then, the solution prepared was coated onto FTO glass surface by brush printing method. After that, the FTO glass was heated at 400°C for 25 min to obtain counter electrode.

Sealing of the Cell

The dye-absorbed TiO$_2$ electrode and Pt counter electrode were attached into a sandwich-type arrangement and sealed with a hot-melted Surlyn film (55 µm, Solaronix) on a hot press. The liquid electrolyte used was prepared by mixing 0.1 M LiI, 0.01 M I$_2$, 0.5 M TBP, and 0.6 M 1,3-dimethylimidazolium with acetonitrile. A drop of the electrolyte was injected into the hole in the back of the counter electrode. Then, the hole was sealed using a cover of low-density polyethylene film (35 µm, Du-Pont). Finally, it is covered by a glass of 0.1 mm thickness. After cleaning the FTO glass outside of the cell, Ag paste was spread onto each side of the cell for electrical connection.

Characterization

The phases present in as received TiO$_2$ powders were identified by powder X-ray diffraction (XRD) analysis. A Rigaku-Geigerflex DMAK/B X-ray diffractometer equipped with Cu target Kα radiation as X-ray source was used to get XRD patterns of the powders. Each powder was scanned in a 2θ range of 20° - 80° at a rate of 2°min-1 by 0.02° increments continuously with an accelerating voltage and applied current 40 kV and 40 mA, respectively. The morphology of TiO$_2$ layer was examined using a scanning electron microscopy (Nova NANOSEM 430). The photocurrent-photovoltage parameters of the prepared cells were measured using a solar simulator (Polaronix K3000, McScience) under illumination of 1000 W/m^2 (AM 1.5). An I-V curve was plotted from which the open-circuit voltage V$_{oc}$ (V), short-circuit current density J$_{sc}$ (mA/cm^2), fill factor (FF), and conversion efficiency η (%) were further attained.

RESULTS AND DİSCUSSİON

X-Ray Diffraction Analysis (XRD)

The XRD analysis revealed that anatase and rutile phases are present in the as received TiO$_2$ powders. The XRD patterns for the powders indicate the characteristic (101) diffraction peak of anatase TiO$_2$ at 2θ of ~25.3° (JCPDS #21-1272) and the characteristic (110) diffraction peak of rutile TiO$_2$ at 2θ of ~27.4° (JCPDS #21-1276) as shown in Figure 1. Quantitative analysis revealed that the powders possess approximately 80 % anatase and 20 % rutile. The XRD pattern of the pastes was similar to the pattern of the as received powders, implying that no new phase and no phase change occurred in the powders after the preparation of paste. A representative XRD pattern for Paste D is shown in Figure 2.

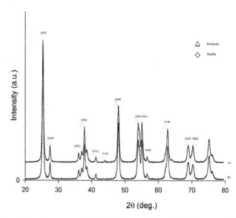

Figure 1. XRD pattern for as received TiO$_2$ powder. (a) 20 nm and (b) 200 nm.

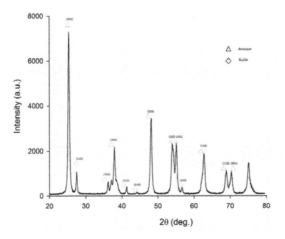

Figure 2. XRD pattern for Paste D.

Scanning Electron Microscopy (SEM)

The surface morphologies of the pastes are shown in Figure 3. The SEM analysis confirms that the presence of small and large particles together causes bigger, interconnected porosity that facilitate dye-loading. (As we didn't measure the dye absorption, increase in dye loading is just assumed) A comparison between the surface morphologies of Paste A and Paste D reveals that TiO$_2$ particles in Paste A are distributed uniformly and size of the largest porosity among the particles are roughly 50 nm as seen in Figure 4 but, the particles in Paste D are not distributed uniformly and porosity as large as 200 nm are apparent as clearly seen in Figure 5.

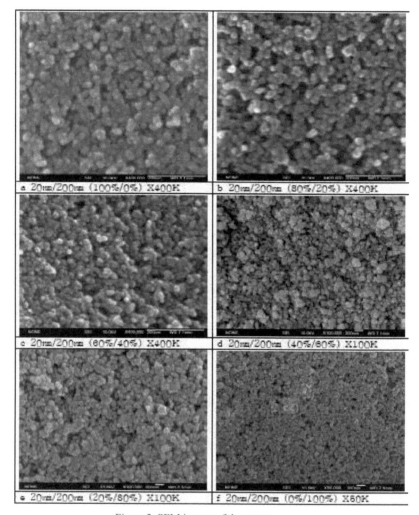

Figure 3. SEM images of the pastes.

Figure 4. SEM image of Paste A.

Figure 5. SEM image of Paste D.

Photovoltaic Measurement of the Cells

Figure 6 shows the photocurrent density-photovoltage characteristics of the DSSCs prepared using different TiO$_2$ pastes. The photocurrent density of the DSSCs ranged between 5 and 13 mA depending on the ratio of small and large TiO$_2$ particles in the pastes. However, the voltage was around 0.8 V for all DSSCs.

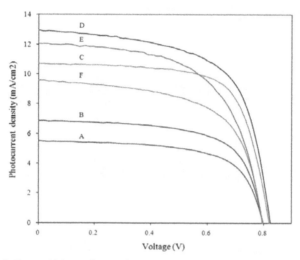

Figure 6. Current-Voltage diagram for the DSSCs prepared using different TiO$_2$ pastes.

The photovoltaic properties; J_{sc}, V_{oc}, FF, and η were calculated from the I-V curves for each DSSC. The values for the photovoltaic properties are listed in Table 2. The conversion efficiency of the DSSC prepared using Paste A (only 20 nm TiO$_2$ particles) was 2.83% which was the lowest value among all DSSCs prepared. The DSSC prepared using Paste F (only 200 nm TiO$_2$ particles) had a η of 4.74%. The η increased by adding 200 nm TiO$_2$ particles to 20 nm TiO$_2$ particles. The highest η value of 6.74% was obtained for Paste D which was composed of 40 wt% 25 nm and 60 wt% 250 nm TiO$_2$ particles. The η decreased with further addition of larger TiO$_2$ particles to small TiO$_2$ particles.

It was observed that J_{sc} displayed a similar trend with η for all DSSCs. The current density increased as the proportion of large particles in the paste was increased. Paste D offered the highest J_{sc} value of 12.85 mA. However, the values of V_{oc} and FF, 0.82 and 0.64, respectively for Paste D were not much different than the values for the other pastes prepared. The increasing trend in J_{sc} and η relative to the increased proportion of larger particles could be a result of better dye absorption characteristics caused by large interconnected porosity in the mixed TiO$_2$ layer.[15] The increase in the volume of porosity in pastes leads to an increase in dye absorption capability. More dye absorption influence by the presence of larger contact points for easier access of the dye can increase the generation of electron-hole pairs. This is evident in the J_{sc} values that increased with increasing proportion of larger particles. The change in J_{sc} was the primary factor in the change of η. In addition, incident light is scattered with larger TiO$_2$ particles, while it penetrates with smaller particles in DSSC.[14] There is a limitation in the improvement in η due to porosity. In the TiO$_2$ paste prepared by using two different size of TiO$_2$ particles, the larger particles makes light scattering easy and increase light harvesting efficiency in the cells. However, as the amount of light scattering particles exceeds a critical value, some of the TiO$_2$ particles may reflect the light causing a decrease in η.

Table 2. The photovoltaic properties for the DSSCs prepared using different TiO$_2$ pastes.

Paste	J_{sc} (mA/cm^2)	V_{oc} (V)	FF	η (%)
A	5.54	0.80	0.64	2.83
B	6.83	0.80	0.64	3.49
C	10.59	0.81	0.70	6.00
D	12.85	0.82	0.64	6.74
E	12.04	0.81	0.62	6.04
F	9.56	0.80	0.62	4.74

CONCLUSİON

A DSSC is produced by using two different sizes of TiO$_2$ particles. The energy conversion efficiency of DSSC is influenced by the porosity in the electrode layer, and could be improved by increasing the size and amount of porosity among the particles. The DSSC prepared by using hybrid TiO$_2$ particles showed a better efficiency than the ones prepared by using either only small or only large TiO$_2$ particles. The maximum efficiency was obtained using a mixture composed of 40 wt% 20 nm and 60 wt% 200 nm TiO$_2$ particles. The increase in DSSC performance is due to increased dye absorption caused by porous structure and light scattering effects of the TiO$_2$ working electrode layer.

ACKNOWLEDGEMENT

The authors are thankful for the partial financial supports received from Atilim University (Project No: BAP-1011-02), Middle East Technical University (Project No: BAP-03-08-2011-003), and the Small and Medium Enterprises Development Organization of Turkey.

REFERENCES

[1] A. Ofir, S. Dor, L. Grinis, A. Zaban, T. Dittrich, J. Bisquert, Porosity Dependence of Electron Percolation in Nanoporous TiO$_2$ Layers, The Journal of Chemical Physics 128 (2008).

[2] R. Cinnsealach, G. Boschloo, S. N. Rao, and D. Fitzmaurice, Coloured Electrochromic Windows Based on Nanostructured TiO$_2$ Films Modified by Adsorbed Redox Chromophores, Sol. Energy Mater. Sol. Cells, 57, 107 (1999).

[3] B. O'Regan, M. Gratzel, A Low-Cost, High-Efficiency Solar Cell Based on Dye-Sensitized Colloidal TiO$_2$ Films, Nature, 353, 737 (1991).

[4] S. Tirosh, T. Dittrich, A. Ofir, L. Grinis, and A. Zaban, Influence of Ordering in Porous TiO$_2$ Layers on Electron Diffusion, J. Phys. Chem. B 110, 16165 (2006).

[5] M. J. Cass, F. L. Qiu, A. B. Walker, A. C. Fisher, and L. M. Peter, Influence of Grain Morphology on Electron Transport in Dye Sensitized Nanocrystalline Solar Cells, J. Phys. Chem. B 107, 113 (2003).

[6] A. Goetzberger, C. Hebling, and H.W. Schock, Photovoltaic materials, history, status and outlook, Mater. Sci. Eng. R, 40, 1 (2003).

[7] H.J. Snaith, Estimating The Maximum Attainable Efficiency in Dye-Sensitized Solar Cells, Adv. Funct. Mater, 20, 13 (2010).

[8] K. Kalyanasundaram, Photochemical and Photoelectrical Approaches to Energy Conversion, Dye-Sensitized Solar Cells, EPFL Press, p. 17-19, 2010.

[9] S. Lee, I-S. Cho, J. H. Lee, D. H. Kim, D. W. Kim, J. Y. Kim, H. Shin, J.-K. Lee, H. S. Jung, N.-G. Park, K. Kim, M. J. Ko, K. S. Hong, Two-Step Sol−Gel Method-Based TiO$_2$ Nanoparticles with Uniform Morphology and Size for Efficient Photo-Energy Conversion Devices, Chem. Mater., 22, 6 (2010).

[10] M. Gratzel, Photoelectrochemical Cells, Nature 414, 338 (2001).

[11] S. Ito, M. K. Nazeeruddin, S.M. Zakeerudddin, P. Pechy, P. Comte, M. Grazel, T. Mizuno, A. Tanaka, T. Koyanagi, Study of Dye-Sensitized Solar Cells by Scanning Electron Micrograph Observation and Thickness Optimization of Porous TiO$_2$ Electrodes Int. J. Photoenergy 2009, 1 (2009).

[12] J. Ferber, J. Luther, Modeling of Photovoltage and Photocurrent in Dye-Sensitized Titanium Dioxide Solar Cells, Sol. Energy Mater. & Sol. Cells 54, 265 (1998).

[13] S. Hore, C. Vetter, R. Kern, H. Smit, A. Hinsch, Influence of Scattering Layers on Efficiency of Dye-Sensitized Solar Cells, Sol. Energy Mater. & Sol. Cells 90, 1176 (2006).

[14]C.-S. Chou, M.-G. Guo, K.-H. Liu, Y.-S. Chen, Preparation of TiO$_2$ Particles and Their Applications in The Light Scattering Layer of a Dye-Sensitized Solar Cell, Appl. Energy 92, 224 (2012).

[15]C.D. Grant, A.M. Schwartzberg, G.P. Smestad, J. Kowalik, L.M. Tolbert, J.Z. Zhang, Characterization of Nanocrystalline and Thin Film TiO$_2$ Solar Cells with Poly(3-undecyl-2,2'-bithiophene) As A Sensitizer and Hole Conductor, J. Electroanal. Chem 522, 40 (2002).

EVALUATION OF COMPRESSION CHARACTERISTICS FOR COMPOSITE-ANTENNA-STRUCTURES

Jinyul Kim[1], Dongseob Kim[2], Dongsik Shin[3], Weesang Park[3], and Woonbong Hwang[1*]
[1] Department of Mechanical Engineering, Pohang University of Science and Technology, San 31, Pohang, Gyungbuk, Republic of Korea
[2] Department of Graduate School of Engineering Mastership, Pohang University of Science and Technology, San 31, Pohang, Gyungbuk, Republic of Korea
[3]Department of Electronic and Electrical Engineering, Pohang University of Science and Technology, San 31, Pohang, Gyungbuk, Republic of Korea
* whwang@postech.ac.kr

ABSTRACT
There are a number of wireless communication systems and broadcasting services which must be integrated into vehicles. Through the innovative integration of antenna elements, amplifiers and ground plane, the reception quality and manufacturability of vehicles is expected to be significantly improved. The most important outstanding problem that is structurally effective materials cannot be used without reducing antenna efficiency. The present study aims to suggest electrically and structurally effective antenna structures which are termed composite-antenna-structures (CAS), and study mechanical and electrical behavior characteristics of CAS after compression test. The CAS is composed of two composite laminates (GFRP, CFRP), nomex honeycomb, and antenna element. And that is the dual-mode annular ring antenna and the microstrip patch antenna for the global positioning system (GPS), satellite digital multimedia broadcasting (DMB) and direct broadcast satellite (DBS). Also, when it is designed, the CAS considered composite materials and adhesive film because of the resonant frequency shift and reduced gain. The GPS antenna patch is designed at the resonant frequency of 1.575GHz, and the peak gain is 6.75dBi. The resonant frequency of DMB is 2.645 GHz, and the gain is 5.6dBi. The resonant frequency of DBS is 12.2 GHz, and the gain is 16.98dBi. The CAS was subjected to compression by using a MTS 810 test facility. When compression load is applied to the CAS, the maximum load was about 12kN and buckling was occurred in GFRP facesheet. The antenna performance of the return loss and radiation pattern was excellent after compression test.

INTRODUCTION
Modern communication systems have experienced high market demand and the trend is likely to continue in the future. As a result of the rapid progress in new wireless communication standards such as GPS, DMB, DBS and new multimedia applications, the demand of wireless terminals capable to operate with multi or wide band is increasing [1-3]. Furthermore, these antennas cannot accommodate multiple antennas due to space, size, weight and cost constraints. There is a conceptual model of the Delphi multiple antenna reception system[4]. The use of a composite roof structure provides a possibility to satisfy the communication and entertainment reception needs with a lightweight and durable self-contained structure. This approach eliminates the multiple mast antennas currently used for AM/FM, television, cellular, and GPS reception. Through the innovative integration of antenna elements, amplifiers and the ground plane, the reception quality and manufacturability of vehicles is improved significantly. In the present study, a new design concept of antennas integrated composite sandwich structures is proposed, as shown in Figure 1. This is to design an electrically and structurally effective multi-functional antenna structure for GPS, DMB and DBS. We designed two types of antennas. One is annular ring patch antenna for GPS and DMB. The other is the microstrip patch array antenna for DBS. Also, the CAS considered the coupling

between antennas, because of considering the interference of 3 bands in one substrate. An attempt has been made to study the buckling characteristics of asymmetric sandwich structure subject to edge compression, when it is applied in vehicles such as bus, train and airplane. The electrical and mechanical performance of test specimens are measured.

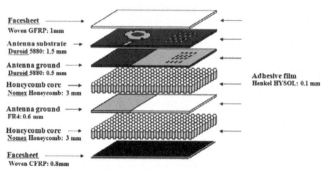

Figure 1. Concept of CAS considering 3 bands.

DESIGN PROCEDURES

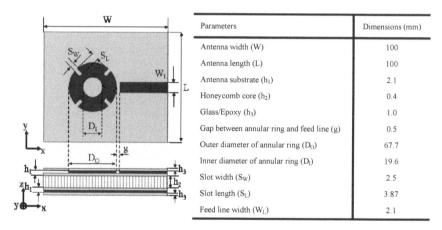

Parameters	Dimensions (mm)
Antenna width (W)	100
Antenna length (L)	100
Antenna substrate (h_1)	2.1
Honeycomb core (h_2)	0.4
Glass/Epoxy (h_3)	1.0
Gap between annular ring and feed line (g)	0.5
Outer diameter of annular ring (D_O)	67.7
Inner diameter of annular ring (D_I)	19.6
Slot width (S_W)	2.5
Slot length (S_L)	3.87
Feed line width (W_L)	2.1

Figure 2. Dimensions of GPS/DMB antenna.

Antenna design

The aims of CSA were to design an antenna for three resonant frequencies at 1.575 GHz (GPS), 2.63 GHz (S-DMB) and 11.7-13.5 GHz(DBS). Two types of antenna structures are proposed; one is an annular ring type and the other is a microstrip patch type. The GPS/DMB antenna comprises an annular ring patch with four slots and a coupling feed line and annular ring patch that had four slots. The resulting antenna elements and their dimensions are shown in Figure 2. To excite two resonance modes, the gap between the feed line and the annular ring patch was used. The resonant frequencies of GPS and DMB were designed by adjusting the inner and outer diameters of the annular ring and by making the four slots in the patch, respectively [5]. And, the DBS antenna consists of microstrip patch and a slot between the antenna substrate and the antenna ground as shown in Figure 3. It used the aperture-coupled feed that the upper substrate can be of low dielectric constant to promote radiation and a lower substrate containing the feed can be of high dielectric constant to enhance binding of the fields to the feed lines. It leads to increased bandwidth. Array antenna can be interconnected to produce a directional radiation pattern and high gain. The antenna were organized into 4×4 array at internals of 0.25λ(6.25mm) for the gain over 10dB.

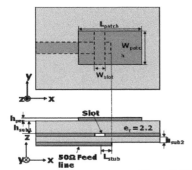

Parameters	Dimensions (mm)
L_{patch}	4
W_{patch}	4.357
W_{slot}	1
L_{stub}	0.75
h_{pec}	0.05
h_{sub1}	1.6
h_{sub2}	0.5
e_r	2.2

Figure 3. Dimensions of DBS antenna.

Antenna coupling
    It is important that the coupling between each band of GPS, DMB, DBS Coupling between the annular ring and array antenna should be minimized for integrating these two antennas on one substrate(20×10cm²). Generally, coupling can be reduced by widening the space between the two antennas. However, if the space widens, coupling reduces the coupling by arranging vertically the each polarization of the two antennas. Two antennas are fabricated in one substrate and the arrows mean the directions of polarization. The measured and simulated S21 are compared to verify the coupling and these results are less than -40 dB. The coupling is minimized from S21.

CAS design
    The CAS was a sandwich panel with inserted antenna elements for signals in the GPS, S-DMB and DBS bands. As shown in Figure 1, this was composed of a GFRP facesheet, antenna substrate, antenna ground, honeycomb core, antenna ground, honeycomb core and CFRP facesheet. The adhesive that bonded the different layers and provided mechanical strength also influenced the antenna performance. The upper facesheet was composed of an woven glass/epoxy laminate that permitted both radiating and receiving radio waves as the antenna functions. The lower facesheet was of woven carbon/epoxy laminate. These facesheets could carry a significant portion of the in-plane mechanical loads, providing resistance to compressive load and environmental factors. The radiating patch layer between the upper facesheet and the antenna ground had a flame retardant composition and incorporated an annular ring type for GPS/S-DMB operation and microstrip patch antenna for DBS operation. The honeycomb core (Showa, CG-SAH 1/8-3.0) was used to support various loads and provide an air-gap for good antenna performance. The ground plane was Duroid 5880 (Rosers Corp., Duroid 5880) and FR-4 (Plastics International, FR4 Glass/Epoxy) plate that

served as the antenna ground. An adhesive film was inserted between each pair of layers to bond the different materials and provide mechanical strength. Table 1 lists the mechanical and electrical properties of the materials used in the CAS. The assembly, covered by a vacuum bag, was then cured in an autoclave according to the recommended curing cycle for this adhesive (125 C for 90 min under a pressure of 2 kg/cm^2).

Table I. Materials and properties.

Materials	Properties	
Composite laminate (Woven GFRP)	Elastic modulus:	25.4 GPa
	Tensile strength:	573.6 MPa
	Dielectric constant:	4.0
	Loss tangent:	0.03
Composite laminate (Woven CFRP)	Elastic modulus:	58.1GPa
	Tensile strength:	716.3MPa
	Conductivity:	1×10^6 s/m
Honeycomb core (Nomex Honeycomb)	Compressive modulus:	414 MPa
	Compressive strength:	7.76 MPa
	Shear strength:	88.6 MPa
	Dielectric constant:	1.1
	Loss tangent:	$\fallingdotseq 0$
Dielectric substrate (Duroid 5880)	Dielectric constant:	2.2
	Loss tangent:	0.001
Dielectric substrate (FR4)	Dielectric constant:	4.5
	Loss tangent:	0.02
Adhesive film (Epoxy Polymer)	Dielectric constant:	2.87
	Loss tangent:	0.028

EXPERIMENTAL

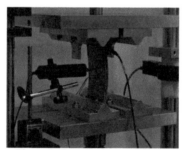

Figure 4. Compression test of CAS

Three specimens were tested using MTS 810 test machine under compressive loadings with a constant crosshead speed 0.5mm/min. The compressive behavior of the CAS specimens was experimentally investigated by applying compressive load in the edgewise direction along the length of the panels as schematically shown in Figure 4. The edgewise compressive test was performed in accordance with the requirements of ASTM C364 with lateral support of the panel facings adjacent to the loaded ends for the simply supported boundary condition. LVDT (linear variable differential transformer) was used to measure the side displacement. The side strains of composite laminates detected through the strain gage.

RESULTS AND DISCUSSION

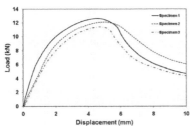

Figure 5. Load - Displacement curve.

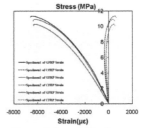

Figure 6. Stress - Strain curve.

Table II Test results.

Specimens	Buckling load (kN)	CFRP strain (με)	GFRP strain (με)	Load (kN)
Specimen 1	7.0	364	6,612	12.63
Specimen 2	7.9	347	6,212	12.11
Specimen 3	6.9	213	6,242	11.40
Average	7.3	308	6,355	12.05

In Table 2 all data obtained from the experiments are listed. The specimens after compression loading returns to their original shape before failure. Therefore, the integrity of the specimens after compression test is maintained. Figure 5 shows the deflection according to apply loads which is the yield and the ultimate load. The average of the peak load is 12kN for the three specimens. As shown in Figure 6, the compressive strain is 6,355με in the GFRP side, and CFRP strain is 308με for the tension. Because the CFRP stiffness is bigger than GFRP, buckling occurred in GFRP side. The CFRP side of specimen subject to load is continuously compressive behavior until 6.3MPa(7.3kN), and it is changed from compression to tension behavior.

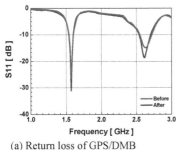

(a) Return loss of GPS/DMB

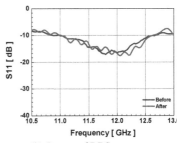

(b) Return of DBS

Figure 7. Return loss of bands.

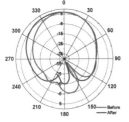

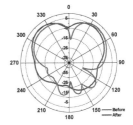

(a)Radiation pattern of GPS(1.575GHz)   (b) Radiation pattern of DMB(2.645GHz)

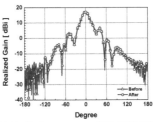

(c) Radiation pattern of DBS(12.2GHz)
Figure 8. Radiation pattern of bands.

The measured electrical performance of the fabricated structures reveals that all reinforcement antennas shifted resonant frequencies and reduced gain. A change in resonant frequency implies a change in the effective dielectric constant of the CAS, and the gain reduces due to loss by the dielectric material in the reinforcement layers. We observed a consistent downward shift in the resonant frequency, by about 0.145 GHz in the GPS range, 0.26 GHz in the S-DMB range and 0.7 GHz in the DBS range, implying a corresponding increase the effective dielectric constant by the reinforcement layers. Figure 7 shows the measured reflection coefficient (S11) where Before and After mean the results of the before and after compression test. The results of S11 satisfy the -10 dB bandwidth in GPS, DMB, and DBS band. Figure 8 shows the radiation pattern of GPS, DMB and DBS. The difference of maximum gains before and after test is small, but the radiation patterns for GPS and DMB band are tilted slightly. However, these results have no influence on the antenna performance. The gains before test are 6.75, 6.25, and 17.15 dBi for GPS, DMB and DBS band respectively and the gain after test are 6.75, 5.62, 16.98 dBi respectively.

CONCLUSIONS
This study involved the design and fabrication of a structurally integrated CAS which is on one substrate ($20 \times 10 cm^2$) for the resonant frequencies of GPS (1.575 GHz), DMB (2.63 GHz) and DMB (11.7~13.5 GHz) considering the coupling of antennas. The annular ring type antenna is available in two frequency bands, which depend on the critical features at the feeding point and in the resonant areas, and the microstrip patch antennas is available in one band which uses the 4x4 array for producing a directional radiation pattern and high gain. For the application of modern transport vehicles, the materials used in the CAS verified corrections for the effect of their composite materials and adhesive films on their electrical performance. Also, it was investigated by

compression loads for its mechanical and electrical behavior. The experimental results show that the preservation of the radiation pattern is strongly dependent on the protection of the antenna elements by composite laminates. The results suggest that even after some local damage, an antenna with composite laminates will continue to function properly in terms of electrical performance.

REFERENCES

[1]F. R. Hsiao, H. T. Chiou, G.T. Lee and K. L. Wong, A Dual-band Planar Inverted-F Patch Antenna with a Branch-line Slit. *Microwave and Optical Technology Letters,* 32 (4), 310-2 (2002)

[2]R. Hossa, A. Byndas, and M. E. Bialkowski, Improvement of Compact Terminal Antenna Performance by Incorporating Open-end Slots in Ground Plane, *IEEE Microwave Wireless Comp. Lett.,* 14(6), 283-5 (2004)

[3]Y. X. Guo, and H. S. Tan, New Compact Six-band Internal Antenna, *IEEE antennas wireless propagate. lett,* 3, 295-7 (2004)

[4]Delphi fuba multiple antenna reception system, Available from: http://www.delphi.com.

[5]J. H. Kim, D. W. Woo, and W. S. Park, Design of a Dual-mode Annular Ring Antenna with a Coupling Feed, *Microwave and Optical Technology Letters,* 51(12), 3011-3 (2009)

DESIGN AND FABRICATION OF SMART-SKIN STRUCTURES WITH A SPIRAL ANTENNA

Dongseob Kim[1], Jinyul Kim[2] and Woonbong Hwang[2*]
[1] Department of Graduate School of Engineering Mastership, Pohang University of Science and Technology, San 31, Pohang, Gyungbuk, Republic of Korea
[2] Department of Mechanical Engineering, Pohang University of Science and Technology, San 31, Pohang, Gyungbuk, Republic of Korea
* whwang@postech.ac.kr

ABSTRACT

We study a composite-antenna-structure (CAS) having high electrical and mechanical performances that we have designed and fabricated. The CAS, consisting of a glass/epoxy face sheet and a honeycomb core, acts as a basic mechanical structure, in which a spiral antenna type is embedded. To increase the intensity, a carbon fiber plate is used as a bottom sheet. This structure of the 0.5 ~ 2 GHz band has a gain of 5 ~ 9 dBi with circular polarization characteristics and reflection loss below -10 dB within the desired frequency band.

INTRODUCTION

In the last 15 years there has been much research into the embedding of antennas in load-bearing structural surfaces of aircraft, so as to improve both structural efficiency and antenna performance [1-3]. Structural, material and antenna designers have collaborated to develop a novel high-payoff technology known as a Conformal Load-bearing Antenna Structure (CLAS) [3]. This technology shows great promise for enhancing the performance and capability of aircraft, by reducing weight, improving the structural efficiency of airframes that contain antennas, and improving the electromagnetic performance of antennas. To develop the load-bearing antenna structure, we proposed the use of antenna-integrated composite structures of sandwich construction, specifically the surface-antenna-structure (SAS) [4-6] and the composite-smart-structure (CSS) [7,8]. In those studies, we designed and fabricated a microstrip antenna structure which implemented satellite communication in the X band (8.2~12.4 GHz) and the Ku band (12.4~18 GHz). At such high frequencies the microstrip antenna has only a small bandwidth. In the present paper we report a new CAS based on a spiral antenna type giving good performance in a low frequency band with higher bandwidth. A sandwich composite consisting of a glass/epoxy face sheet and honeycomb core is used as a basic mechanical structure, in which a spiral antenna type is embedded. To increase the intensity, a carbon fiber plate is used as the bottom sheet.

DESIGN PROCEDURES

The basic design concept of the CAS panel is an organic composite multi-layer sandwich panel into which spiral antenna elements are inserted. This concept originates mechanically from a composite sandwich structure, and electrically from a spiral antenna, as shown in Figure 1. The sandwich structure consists of two thin load-bearing facesheets, bonded to either side of a moderately thick and lightweight core that prevents the face sheets from buckling. The sandwich structure gains its bending rigidity mainly by separating the facesheets, and has very high structural efficiency (ratio of strength or stiffness to weight). The SAS panel consists of several basic layers. Each layer must meet its own combination of structural and electrical design requirements, as well as the manufacturing and assembly requirements. The basic panel layers are: an outer facesheet, antenna element, honeycomb core, and supporter elements. These are shown in figure 2 in an exploded view, which also specifies the materials chosen in each layer. The layers are bonded by adhesive to form the final assembly. The outer facesheet must carry a significant portion of the in-plane loads, since it contributes to the overall panel buckling resistance, and it also provides low velocity impact and environmental resistance.

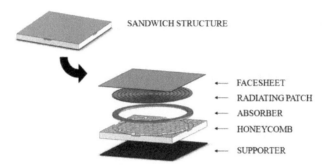

Figure 1 Design concept of the composite antenna structure

This outer facesheet must also permit the transmitting and receiving of RF signals. The facesheet material must be low loss and only weakly dielectric in order to minimize signal attenuation and reflection loss. The honeycomb cores transmit shear loads between layers induced by bending loads in the panel, and support the facesheet against compression wrinkling. They also provide impact resistance and increase the overall panel buckling resistance. The thickness of the honeycomb cores contributes significantly to the overall rigidity, and is involved in the balance between panel thinness and structural rigidity.

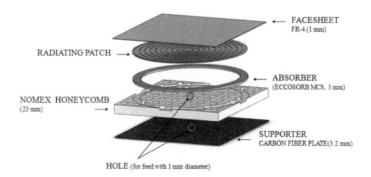

Figure 2 Structure and materials of the composite antenna structure

The supporter also carries a significant portion of the in-plane loads together with the outer facesheet, as well as supporting the whole structure. It can be selected without need to consider electrical performances, and therefore has the best mechanical properties of any layer in the CAS construction. Spiral antennas [10, 11] can be used in high-performance aircraft, spacecraft, and in satellite and missile applications, where constraints include size, weight, cost, performance, ease of installation, and aerodynamic profile. These antennas are low-profile, conformable to planar and nonplanar surfaces, simple and inexpensive to manufacture using modern printed-circuit technology, and compatible with MMIC designs. Our CAS design is based on a spiral antenna type with a bottom layer of carbon fiber plate. The antenna does not work well if the the spiral antenna current is interrupted. When the reflector is located near the spiral antenna, the currents on the spiral arms are not reduced enough at the end of arms. There are some reflections at the end of spiral arms. In other words, the reflector interrupts radiations and the wave from the antenna source flows to the end of the arms without enough radiations. To overcome these problems, we used an electromagnetic wave absorber on a hole in honeycomb core. The spiral antenna is placed on the absorber, which absorbs electromagnetic waves. Unwanted radiation cannot pass through the absorber, reducing the effect of the reflector. If we do not use the absorber, there are some reflections at the end of the spiral arms. This reflection distorts not only the radiation patterns but also polarizations. We cannot obtain wideband axial ratio performance without the absorber.

DESIGN AND EXPERIMENTAL PROCEDURES

The antenna is to be designed for low frequency and broadband communication. The antenna requirements are: frequency range 0.5 to 2 GHz (bandwidth 1.5 GHz), and gain at least 10 dBi with circular polarization. In designing the antenna elements, a computer-aided design tool (CST Microwave Studio) is used to select a large number of strongly interacting parameters by means of integrated full-wave electromagnetic simulation. The resulting antenna elements and their dimensions are shown in Figure 3.

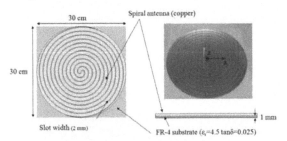

Figure 3 Design of the spiral antenna element

The facesheet is used to a FR-4 glass/epoxy radiating patch comprising a spiral antenna, 1 mm thick. This spiral antenna, set in a circle of diameter 30 cm, is 2 mm thick and has 10.5-turns. FR-4 glass epoxy is a popular and versatile high pressure thermoset plastic laminate grade with good strength to weight ratio. FR-4 undergoes negligible water absorption and is commonly used as an electrical insulator possessing considerable mechanical strength. The main objective is to obtain a good impedance match as seen by the feedline, in the range of frequencies from 0.5 to 2 GHz. The feedline connected to the input port has characteristic impedance 50 Ω, chosen for impedance matching at the port. Coaxial cable is used to feed the antenna in the center. Manufacture of the CAS is a sequential process. The facesheet, including antenna elements, are first prepared by a photolithographic process. The honeycomb cores and each layer must be aligned prior to permanent bonding, in order to give precise electromagnetic coupling. For alignment, four guide holes are made near the edge of all layers. These are confirmed to have no effect on antenna performance. The CAS is assembled by aligning these holes using a plastic nut and bolt. Each layer is bonded to the top and bottom of its neighbors in the designed sequence, using epoxy film adhesive. The assembly, covered by a vacuum bag, is then cured in an autoclave according to the recommended curing cycle for this adhesive (125 °C for 90 minutes at a pressure of 3 kg/cm2).

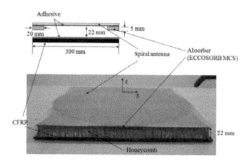

Figure 4 Structure and fabrication of the composite antenna structures

Figure 4 shows the appearance of each layer and the top view of the final assembly after fabrication. The size of the CAS is 300 × 300 × 27 mm. Antenna performance of the fabricated CAS is determined by electrical measurements. The return loss characteristic, which measures the mismatch or the ratio of the reflected power to the incident power at the input port, is measured using a Network Analyzer 8510 under laboratory conditions. The radiation patterns are measured in an anechoic chamber at four frequencies, 0.5, 1, 1.5 and 2 GHz, in order to show the bandwidth patterns. Gains and axial ratios are calculated by comparing the magnitude of the electric field against a standard-gain horn antenna.

RESULTS AND DISCUSSIONS

We have studied an antenna embedded in a structural surface, which provides good structural and good electrical efficiencies at the same time. The design is a composite sandwich structure in which a spiral antenna element has been inserted. This design provides antenna performances that meet our requirements.

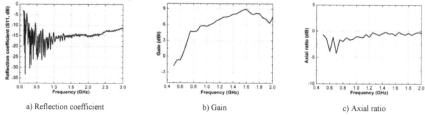

a) Reflection coefficient   b) Gain   c) Axial ratio

Figure 5 Electrical performance of composite antenna structure

Figure 5 shows the electrical performance of the composite antenna structure. Figure 5(a) shows the return loss characteristic; a bandwidth of approximately 1.5 GHz is seen, corresponding to the frequency range of interest (0.5-2 GHz). Figure 5(b) shows the antenna gain and the gain reduction occurs at low frequencies and is caused energy loss due to absorption. Figure 5(c) shows the axial ratio, and measuring less than 3 dB within the band and get the value of circular polarization is well formed.

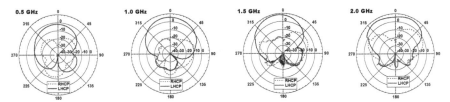

Figure 6 Radiation patterns in bandwidth

Figure 6 shows the radiation patterns at 0.5, 1, 1.5 and 2 GHz. The radiation pattern indicates that emissions from the front of the spiral antenna is LHCP (Left Hand Circularly Polarized), RHCP (Right Hand Circularly Polarized) include coming out the back. The back radiation is low because of the carbon fiber plate, and front radiation is large with a LHCP characteristic.

CONCLUSIONS

In this paper, we have designed and fabricated a spiral antenna of honeycomb sandwich construction. The final demonstration article is a 300 × 300 × 27 mm flat antenna panel with an antenna element. Electrical measurements of the fabricated structure show that is satisfies the design requirements, with a bandwidth above 1.5 GHz and a high gain with circular polarization. In addition, it has been based on a carbon fiber plate, so this structure has a higher intensity than other structure with a glass fiber plate. The design concept can be extended to give a useful guide for manufacturers of structural body panels as well as antenna designers, and promises to be an innovative future communication technology.

ACKNOWLEDGEMENT

This research was financially supported by the Ministry of Education, Science Technology (MEST) and Korea Institute for Advancement of Technology(KIAT) through the Human Resource Training Project for Regional Innovation. This work was supported by the Brain Korea 21 Program administrated by the Ministry of Education of Korea.

REFERENCES

[1]A.J. Lockyer, K.H. Alt, D.P. Coughlin, M.D. Durham, and J.N. Kudva, Design and development of a conformal load-bearing smart-skin antenna: overview of the AFRL Smart Skin Structures Technology Demonstration (S3TD), *Proceedings of SPIE*, 3674 (1999)

[2]A.J. Lockyer, J.N. Kudva, D.Kane, B.P Hill, and C.A Martin: *Proceedings of SPIE*, 2189 (1994)

[3]J. Tuss, A.J. Lockyer, K. Alt, F. Uldirich, R. Kinslow, J. Kudva and A. Goetz, Qualitative assessment of smart skins and avionics/structures integration, *37th AIAA Structural Dynamics and Materials Conference*, 1415

[4]C.S. You, W. Whang, Park H.C., Lee R.M. and Park W.S., Microstrip antenna for SAR application with composite sandwich construction: surface-antenna-structure demonstration., *Journal of Compos Materials*, 37(4)   (2003)

[5]J.H. Jeon, W. Hwang, Park H.C. and Park W.S., Buckling characteristics of smart skin structures., *Composite Structures*, 63(3-4) (2004).

[6]D.H. Kim, W. Hwang, Park H. C. and Park W. S., Fatigue characteristics of a surface antenna structure designed for satellite communication., *Journal of Reinforced Plastics and Composites*, 24(1) (2005)

[7]C.S. You, W. Hwang and Eom S.Y., Design and fabrication of composite smart structures for communication, using structural resonance of radiated field., *Smart Materials and Structures*, 14 (2005)

[8]C.S. You and W. Hwang, Design of load-bearing antenna structures by embedding technology of microstrip antenna in composite sandwich structure., *Composite Structures*, 71 (2005)

Advanced Ceramics and Composites for Sustainable Nuclear Energy and Fusion Energy

COMPARISON OF PROBABILISTIC FAILURE ANALYSIS FOR HYBRID WOUND
COMPOSITE CERAMIC ASSEMBLY TESTED BY VARIOUS METHODS

James G. Hemrick and Edgar Lara-Curzio
Material Science and Technology Division, Oak Ridge National Laboratory
Oak Ridge, TN, USA

ABSTRACT
    Advanced ceramic matrix composites based on silicon carbide (SiC) are being considered
as candidate material systems for nuclear fuel cladding in light water reactors. The SiC
composite structure is considered because of its assumed exceptional performance under
accident scenarios, where its excellent high-temperature strength and slow reaction kinetics with
steam and associated mitigated hydrogen production are desirable. Among the specific structures
of interest are a monolithic SiC cylinder surrounded by interphase-coated SiC woven fibers in a
tubular form and infiltrated with SiC. Additional SiC coatings on the outermost surface of the
assembly are also being considered to prevent hydrothermal corrosion of the fibrous structure.
The inner monolithic cylinder is expected to provide a hermetic seal to contain fission products
under normal conditions. While this approach offers the promise of higher burn-up rates and
safer behavior in the case of loss-of-cooling accident (LOCA) events, the reliability of such
structures must be demonstrated in advance. Therefore, a probability failure analysis study was
performed of such monolithic-composite hybrid structures to determine the feasibility of these
design concepts. This analysis can be used to predict the future performance of candidate
systems in an effort to determine the feasibility of these design concepts and to make future
recommendations regarding materials selection.

INTRODUCTION
    Recent interest in advanced ceramics and ceramic matrix composite assemblies based on
silicon carbide (SiC) for nuclear fuel cladding applications in light water reactors (LWRs) has
led to the initiation of a test effort at Oak Ridge National Laboratory (ORNL) to characterize the
strength of these materials. SiC is known to possess high strength, reasonable thermal
conductivity, and low chemical reactivity at high temperatures. These combined properties are
especially attractive when compared to those of traditionally used zirconium alloys such as
Zircaloy™. It is therefore expected that based on these properties, SiC composite assemblies
could act as nuclear fuel cladding which would remain intact and safe during and after long
periods of time at very high temperatures, such as those occurring in loss-of-cooling accident
(LOCA) events at nuclear reactors or spent fuel pools. Additionally, the low chemical reactivity
of SiC is expected to mitigate oxidizing reactions that break down water molecules producing
free hydrogen gas that is not only potentially explosive, but may embrittle tubes (such as the
formation of zirconium oxide and hydride in zirconium alloys).
    One of the proposed concepts for SiC-based fuel cladding is a monolithic SiC cylinder
surrounded by carbon-coated SiC fibers woven into a tubular form and infiltrated (by chemical
vapor deposition) with SiC. Additional SiC coatings on the outermost surface of the assembly
are also being considered to prevent hydrothermal corrosion of the fibrous structure [1]. In such a
structure, the inner monolithic SiC cylinder is expected to provide a hermetic seal to contain the
fuel pellets and fission products. The composite structure is designed to provide high strength,
high stiffness, and high toughness as the ceramic matrix transfers stress to the fibers, while the
carbon interface layer on the fibers provides a mechanism for crack deflection to enable a
graceful mode of failure. While such an approach offers the promise of higher burn-up rates and
safer behavior in the case of LOCA events, the reliability of such ceramic-based structures still

needs to be demonstrated in advance, in particular its ability to reliably retain fission products under normal operation. Therefore, a series of probability failure analysis studies were undertaken at ORNL to consider the feasibility of these design concepts.

BACKGROUND

A previous study [2] was performed based on applying known loading conditions such as axial loads, vibrational loads, and temperature differential across the thickness of a hybrid ceramic assembly and along its length to define a hypothetical stress state for a characteristic fuel rod and the associated cladding system. Using such a stress state, a probabilistic analysis was performed to determine the reliability of a candidate system. From this analysis, a range of values of characteristic strength and Weibull modulus of materials in such a hybrid structure that are needed to meet allowable failure rates was identified. Additionally, finite element analysis (FEA) was performed to compute the stress state for the monolithic inner cylinder of a hybrid monolithic-composite SiC fuel cladding system, which is responsible for containing the fuel and fission products based on a tube geometry of outer radius (r_o) and inner radius (r_i) considering a characteristic volume of length (l) for a single event in the loading history of the system as shown in Figure 1.

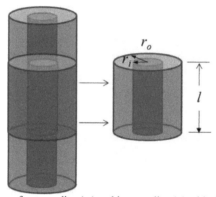

Figure 1. Tube geometry of outer radius (r_o) and inner radius (r_i) with characteristic volume of length (l) considered for FEA.

Existing SiC failure data from the literature was used to analyze probability of failure behavior due to surface and volume flaws as a function of Weibull modulus and characteristic strength as demonstrated hypothetically in Figure 2 where the Weibull modulus (m) and deterministic stress (σ_o) are determined by the slope of each curve. For this analysis, the "size effect" encountered when considering the failure strength of brittle materials and the concept of the "weakest link" where the failure of one link leads to total failure of a component were also considered. Details of this analysis are given elsewhere [3].

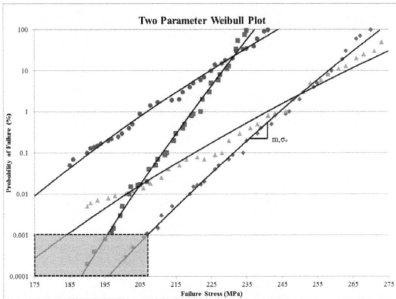

Figure 2. Hypothetical example of experimental failure data used to analyze behavior due to certain flaw population (grey box indicates an allowable probability of failure of 0.001%).

EXPERIMENTAL

Based on the understanding gained from the previously described work concerning the probability of failure for a hypothetical SiC system, work was undertaken to evaluate how to determine failure data for the actual SiC systems of interest. Such data must be experimentally obtained and analyzed in the same manner as above to obtain predictions relevant to the actual candidate material system. Additionally, the test data must be generated in a fashion that yields information relevant to the failure mode of these materials in actual service.

Mechanical test data were desired for both commercially obtained monolithic SiC cylinders and SiC cylinders incorporated into actual woven composite systems as both types of data will be needed for full evaluation of candidate material systems for nuclear fuel cladding. This paper focuses on data generated on monolithic SiC. Additionally, test data generated by multiple test methods were desired in an effort to evaluate how data from various methods compare to one another and to evaluate the relevance of the data from each test in regard to how ceramic fuel clad assemblies are expected to fail in actual service.

The first data considered were conventional tensile test data generated under uniaxial loading. This is a standard test method for determining the strength of tubular materials and was expected to provide base line information for test method comparison and future modeling efforts. Yet, data from this type of testing may not be directly comparable to tensile hoop data that may be more characteristic of the expected failure mode inherent in failed clad tubing due to service operation loading. Therefore, the bulk of testing carried out under this project was generated using an internal pressurization technique developed at ORNL by radially expanding a short ring of the test material through application of an axial compressive load to a cylindrical plug of polyurethane fitted inside the specimen [4]. A schematic of the test setup for this method is shown in Figure 3. For such an internal pressurization condition (P), the axi-symmetric hoop stress ($\sigma(r)$) can be expressed as a function of the radial distance (r) from the centerline of the specimen as shown in Equation 1:

$$\sigma(r) = \frac{r^2 P}{r_o^2 - r_i^2}\left(1 + \frac{r_o^2}{r^2}\right),$$ (1)

where r_i and r_o are the inner and outer radii of the specimen, respectively and the fracture stress (σ_f), which occurs at the maximum load when fracture occurs (P_f), is given as the stress at the inner surface ($r = r_i$) as shown in Equation 2 [5].

$$\sigma_f = \frac{r_o^2 + r_i^2}{r_o^2 - r_i^2} P_f$$ (2)

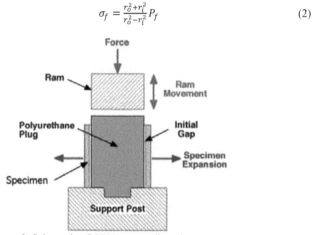

Figure 3. Schematic of ORNL expanding plug test setup.

The final test method employed was a diametral compression test to generate flexure data though crushing. For this test, a tubular specimen was fractured by diametrical compressive loading as shown in Figure 4.

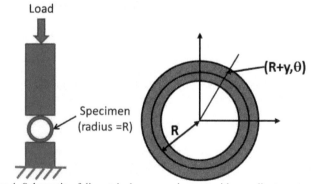

Figure 4. Schematic of diametrical compression test with coordinate system.

The stress distribution for this case is based on curved beam theory [6] where if a tube is subjected to a load per specimen length (L) applied through the centroid axis of the outer surface, the resultant elastic hoop stress is the sum of the stress components caused by an axial load and by a bending load [4]. The bending moment (M) is a function of the angle from the horizontal line (θ):

$$M = \frac{1}{2} LR \left(\frac{2}{\pi(1+z)} - \cos\theta \right), \tag{3}$$

where:

$$R = \frac{1}{2}(r_i + r_o),$$
$$z = \frac{R}{2c} \ln\left(\frac{R+c}{R-c}\right) - 1,$$
$$c = \frac{1}{2}(r_o - r_i).$$

The hoop stress at a point (y,θ) is then given by:

$$\sigma = \frac{L\cos\theta}{2a} + \frac{M}{aR}\left(1 + \frac{1}{z}\frac{y}{R+y}\right), \tag{4}$$

for (a = 2cl) and a specimen of length (l) loaded at a distance (y) from the mid-thickness position of the specimen wall ($-c \leq y \leq c$). The maximum stress will occur at the inner surface of the tube (y = -c, θ = 90°) or at the inside of the point where load is applied (y = +c, θ = 90°). When fracture occurs at a load (L_f), the fracture stress (σ_f) is then given by:

$$\sigma_f = \frac{L_f}{\pi a(1+z)}\left(1 - \frac{1}{z}\frac{c}{R-c}\right). \tag{5}$$

RESULTS AND DISCUSSION

α-SiC tubular material (Hexoloy SE SiC Tube (2012 Vintage) with nominal outer diameter (OD) = 14 mm, inner diameter (ID) = 11, wall thickness = 1.5 mm, length = 254 mm) was obtained from Saint-Gobain Ceramic Materials Structural Ceramics Division (Niagara Falls, NY). This was off the shelf stock material characteristic of commercially available SiC tubing. Test specimens were prepared from four random tubes of this material for tensile, expanded plug and flexure testing as shown in Table 1.

Table 1. Hexoloy SE SiC Test Specimen Dimensions

Tensile Specimens	OD (mm)	Wall (mm)	ID (mm)	Length (mm)
1	13.83	1.54	10.75	101.68
2	13.82	1.50	10.82	101.72
3	13.89	1.51	10.87	101.79
4	13.89	1.54	10.80	101.73
5	13.80	1.55	10.71	101.70
6	13.90	1.54	10.83	101.67
Plug Test Specimens	**OD (mm)**	**Wall (mm)**	**ID (mm)**	**Length (mm)**
1	13.85	1.53	10.78	10.23
2	13.93	1.54	10.86	10.25
3	13.98	1.53	10.92	10.13
4	13.97	1.53	10.91	10.31
5	13.94	1.53	10.88	10.29
6	13.96	1.53	10.89	10.13
7	13.96	1.54	10.89	10.23
8	13.93	1.54	10.84	10.24
9	13.99	1.53	10.92	10.23
10	13.98	1.53	10.92	10.15
11	13.97	1.56	10.85	10.25
12	13.97	1.53	10.90	10.21
13	14.02	1.56	10.90	10.23
14	13.96	1.54	10.87	10. 24
15	13.95	1.54	10.87	10.23
16	13.99	1.58	10.84	10.23
17	13.93	1.53	10.87	10.16
18	13.95	1.53	10.90	10.26
19	14.06	1.53	11.00	10.23
20	13.95	1.56	10.84	10.20

Table 1(Continued)

Flexure Test Specimens	OD (mm)	Wall (mm)	ID (mm)	Length (mm)
1	13.87	1.52	10.82	10.23
2	14.05	1.53	10.99	10.22
3	14.02	1.52	10.99	10.25
4	14.06	1.54	10.98	10.24
5	13.94	1.53	10.88	10.29
6	14.02	1.52	10.99	10.26
7	13.96	1.54	10.89	10.23
8	13.97	1.54	10.89	10.25

Existing tensile test data previously generated at ORNL for Saint-Gobain Enhanced Hexoloy SA (2001 Vintage) material were used as baseline data for this study [7]. Tensile properties and Weibull statistics for this material were generated for a twenty test specimen sample resulting in a characteristic strength of 320 MPa and a Weibull Modulus of 10.0. As indicated in Table 1, test specimens were prepared to validate these results using the current material, but such work had not yet been performed at the time of the writing of this paper.

Table 2. Expanding Plug Test Results for Hexoloy SE SiC

Plug Test Specimen	Axial Failure Load (kN)	Calculated Circumferential Stress (MPa)
1	1.847	64.90
3	3.058	108.52
4	3.090	107.74
5	4.461	155.85
7	3.704	129.31
8	3.055	106.55
9	3.066	107.74
10	4.037	142.96
11	3.493	120.16
12	3.991	140.51
13	2.377	81.93
14	4.436	154.70
15	3.090	107.88
16	3.987	135.67
18	4.243	148.65
19	3.619	127.18
20	3.729	128.91

Avg: 121.71, Stdev: 24.90 MPa

Expanding plug testing was successfully performed on 17 of the 20 test specimens shown in Table 1. Three test specimens were retained from the original sample for future testing or validation efforts. Results of testing are shown in Table 2 with failure loads converted to circumferential stresses based on sample dimensions and previously given equations.

This sample set was then used to generate Weibull statistics for this material under internal pressurization conditions. A Weibull plot is shown in Figure 5. A characteristic strength of 132 MPa and a Weibull Modulus of 5.3 was determined for this material under these loading conditions.

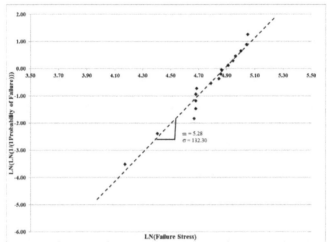

Figure 5. Weibull Plot for Hexoloy SE SiC Under Internal Pressurization Conditions.

Diametral compression flexure testing was successfully performed on five of the eight test specimens shown in Table 1. One test specimen from the original sample set was accidentally broken during loading and two specimens were used for ring expansion testing (test specimen numbers retained and reported in Table 2). Results of testing are shown in Table 3 with failure loads converted to failure stresses based on sample dimensions and previously given equations.

Table 3. Diameteral Compression Flexure Test Results for Hexoloy SE SiC

Test Specimen	Failure Load (kN)	Calculated Failure Stress (MPa)
2	0.415	251.02
3	0.401	245.89
4	0.410	245.91
6	0.430	263.10
8	0.406	241.17

Avg: 249.42, Stdev: 7.52 MPa

This sample set was then used to generate Weibull statistics for this material in diametral compression flexure. A Weibull plot is shown in Figure 6. A characteristic strength of 253 MPa and a Weibull Modulus of 31.7 was determined for this material under these loading conditions. It should be noted that these statistics were generated based on only five tests leading to an upper bound of 46.4 and lower bound of 11.4 on the Weibull Modulus estimate calculated (90% confidence interval).

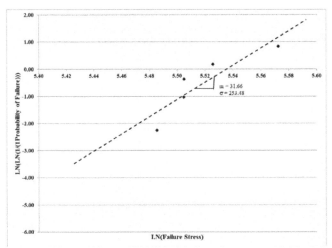

Figure 6. Weibull Plot for Hexoloy SE SiC in Diametral Compression Flexure.

As can be noted from the experimental test data presented above, axial tensile testing resulted in a much higher characteristic strength estimate (320 MPa) than plug (132 MPa) or flexure testing (253 MPa). This is thought to be partly due to the direction of loading introduced by each test method and the inherent flaw populations in these materials that lead to failure. Axial tensile testing is expected to promote failure due to volume distributed flaws, while flexure testing is more prone to promote failure due to surface flaws. The expanded plug testing is expected to sample both flaw populations. Additionally, strength differences between the testing methods can be attributed to the very different effective surfaces/volumes present in the associated test specimens.

Also, each of the above Weibull analyses assumes that only a single flaw population dictates failure in these materials. This is expected to be a false assumption as discussed above and it is probable that both surface and volume flaws contribute to the failure and may be dominant in different test specimens. Therefore, fractography is an important tool that needs to be incorporated into the analysis. Due to the catastrophic failure of the test specimens, fractography of the current test samples proved difficult. Additional testing is currently under way in a fashion to improve retention of test specimens in a more complete manner to aid in fractographic analysis.

Additionally, it is believed that a greater number of test specimens may need to be evaluated to improve the generated Weibull statistics. The ASTM standard applicable to generating Weibull statistics for advanced ceramics is Standard Practice C 1239 [8]. As stated in this document, it is not the number of parameters in the Weibull distribution that is of importance when trying to maximize the accuracy of the predictions (two versus three parameter formulations), but the desired confidence bounds on the predicted distribution parameters. Therefore the desired confidence bounds are what really determine the required sample population size. Additionally, the required population size may be material specific and will most likely not be universal for all systems. Therefore, one may not be able to predict the needed sample population size a priori and may have to perform experiments followed by fractography of the failed specimens to identify the population and location (volume vs. surface) of the strength-limiting flaws as testing progresses to determine the needed sample size based on failure modes and the desired confidence bounds of the predicted distribution parameters. This could be especially important as there could be different distributions of strengths, with each distribution associated with one type of strength limiting flaw. Only fractography can help one to identify the nature and characteristics of those flaws.

Although it is not possible to state a fixed number of test specimens that will need to be tested to produce an accurate Weibull distribution which characterizes candidate ceramic fuel cladding materials, it is felt that this number will be dependent on the individual candidate material system tested and will be further complicated by the complex interaction of the ceramic matrix and fiber reinforcement that make up candidate composite SiC cladding systems. It can be anticipated though that larger sample populations than those used in the current work will be needed. Based on the guidance found in ASTM C 1239, sample populations on the order of at least 50-100 may be required.

Comparison of the experimental data generated from this project were also performed with previous strength data generated on Saint-Gobain Enhanced Hexoloy SA (2001 Vintage) material at ORNL [7]. Testing was performed on this material in flexure using both a c-ring (expected to sample surface flaws perpendicular to the circumferential direction) and sectored flexure test methodology (expected to sample surface flaws parallel to the circumferential direction). C-ring testing of 18 test specimens resulted in a characteristic strength of 299 MPa and Weibull modulus of 4.63. Sectored flexure testing of 20 test specimens resulted in a characteristic strength of 192 MPa and Weibull modulus of 5.5. These values bracket the diametral flexure test results (expected to sample surface flaws both perpendicular and parallel to the circumferential direction) generated under this project (characteristic strength of 253 MPa). These flexure values are greater than the failure stress estimate generated by plug testing (characteristic strength of 132 MPa). Again this is thought to be due to the direction of loading introduced by this test method and the inherent flaw populations sampled by this test method. Since tensile hoop data from this type of testing may be more directly comparable and characteristic of the expected failure mode inherent in failed cladded tubes due to service operation loading it is of note that this test method provides the lowest predicted strength for this material.

CONCLUSION

Traditional tensile testing generated a characteristic strength of 320 MPa and a Weibull Modulus of 10.0 for the standard α-SiC material tested under this project (90% Confidence Interval range on strength estimate – 307 to 334 MPa). A characteristic strength of 132 MPa and a Weibull Modulus of 5.3 was generated by expanded plug testing for this same material (90% Confidence Interval range on strength estimate – 121 to 144 MPa). The third test method employed, diametral compression flexure testing, generated a characteristic strength of 253 MPa and an unrealistically high Weibull Modulus of 31.7 for this material which is thought to be due to the limited number of test specimens (90% Confidence Interval range on strength estimate – 244 to 263 MPa).

As previously observed, axial tensile testing resulted in a much higher failure stress estimate than plug or flexure testing. This is thought to be due to the direction of loading introduced by each test method, the effective surfaces/volumes present in the associated test specimens, and the inherent flaw populations in these materials that lead to failure. Also, each of the above Weibull analyses assumes that only a single flaw population dictates failure in these materials which is expected to be a false assumption. Therefore, fractography is an important tool that needs to be incorporated into this analysis. Since tensile hoop data from this type of testing may be more directly comparable and characteristic of the expected failure mode inherent in failed clad tubing due to service operation loading, it is also of note that this test method yields the lowest predicted strength for this material. Additionally, it is believed that a greater number of test specimens may need to be evaluated to improve the generated Weibull statistics. In the future, it is hoped that data generated from this analysis will be used to predict the future performance of candidate systems in an effort to determine the feasibility of these design concepts and to make future recommendations regarding materials selection.

ACKNOWLEDGEMENTS

The authors wish to thank Dr. Yong Yan for assistance with diametrical compression testing described in this paper and Dr. Andrew Wereszczak for use of archived SiC test data. Work described in this paper was supported by the United States Department of Energy Office of Nuclear Energy under the Light Water Reactor (LWR) Sustainability Program's LWRS Advanced Nuclear Fuels Pathway. This submission was produced by a contractor of the United States Government under contract DE-AC05-00OR22725 with the United States Department of Energy. The United States Government retains, and the publisher, by accepting this submission for publication, acknowledges that the United States Government retains, a nonexclusive, paid-up, irrevocable, worldwide license to publish or reproduce the published form of this submission, or allow others to do so, for United States Government purposes.

REFERENCES

[1] E. Barringer, Z. Faiztompkins, H. Feinroth, T. Allen, M. Lance, H. Meyer, L. Walker, and E. Lara-Curzio, "Corrosion of CVD Silicon Carbide in 5001C Supercritical Water," *J. Am. Ceram. Soc.*, 90, (1) 315–318 (2007).

[2] O. Jadaan, "Feasibility of Using Monolithic Ceramics for Nuclear Fuel Cladding Applications," internal report, December (2011).

[3] J.G Hemrick and E. Lara-Curzio, "Probabilistic Failure Analysis for Wound Composite Ceramic Cladding Assembly", ORNL Technical Report, ORNL/LTR-2012/198, (2012).

[4] W. J. McAfee, W. R. Hendrich, T. E. McGreevy, C. A. Baldwin, and N. H. Packan, "Postirradiation Ductility Demonstration Tests of Weapons-Derived Fuel Cladding," Flaw Evaluation, Service Experience, and Materials for Hydrogen Service, PVP-Vol. 475, July 2004, pp. 213–220 (2004).

[5] T.S. Byun, E. Lara-Curzio, R.A. Lowden, L.L. Snead, and Y. Katoh, "Miniaturized fracture stress tests for thin-walled tubular SiC specimens," Journal of Nuclear Materials, 367-270, 633-658, (2007).

[6] F.B. Seely, J.O. Smith, Chapter 6: Curved flexural members in *Advanced mechanics of materials*, second ed., John Wiley & Sons, Inc., (1961).

[7] J J.Schwab, A. A. Wereszczak, J. Tice, R. Caspe, R. H. Kraft, and J.W. Adams, "Mechanical and Thermal Properties of Advanced Ceramics for Gun Barrel Applications,"*ARL-TR-3417*, February (2005).

[8] "Standard Practice for Reporting Uniaxial Strength Data and Estimating Weibull Distribution Parameters for Advanced Ceramics," ASTM C 1239, Annual Book of ASTM Standards Vol. 15.01, American Society for Testing and Materials, West Conshohocken, PA, (2012).

STRENGTH – FORMULATION CORRELATIONS IN MAGNESIUM PHOSPHATE CEMENTS
FOR NUCLEAR WASTE ENCAPSULATION

W. Montague*, M. Hayes[+] & L.J. Vandeperre*
*Centre of Advanced Structural Ceramics & Department of Materials, Imperial College London, South Kensington Campus, London SW7 2AZ, UK
[+] National Nuclear Laboratory, Warrington WA3 6AE, UK

ABSTRACT
        Nuclear waste streams consist of a diversity of physical and chemical forms, potentially requiring a toolbox approach in the application of materials to provide suitable matrices for their disposal. Magnesium Potassium Phosphate Cement (MKPC) is an alternative to Portland cement for encapsulation of specific wastes where lower free water content, lower internal pH or the immobilisation of radioactive isotopes as low-solubility phosphates minerals are desired properties. The aim of this work was to support mix design through establishing strength-formulation relationships for MKPC blends with pulverised fuel ash (pfa). This is achieved utilising relative density calculations based on measurements of bound water for the determination of the progression of the setting reactions and relating this to compressive strength development data for a variety of formulations. While 40% of the chemical reaction occurs in the first day after mixing, true strength development is found to occur much later as the initial reaction products merely gels the slurry until a critical relative density is reached and strengthening becomes substantial. Measured compressive strength values for the system have reached as high as 70 ± 5 MPa at 180 days of age.

INTRODUCTION
        The UK has a considerable amount of historic nuclear waste, which was generated during the lifetime of existing nuclear power plants and now also during decommissioning of nuclear installations. There is a wide range of wastes including some difficult legacy wastes arising through the historic practice of mixed storage of waste. The inhomogeneous nature of the waste stream is one of the challenges in formulating reliable and robust approaches to waste management in the nuclear sector[1].
        The use of cements to encapsulate and immobilize intermediate level waste (ILW) has a long track record in the UK and evidence of its effectiveness has accrued over the years[2]. Cements are effective encapsulants as they provide a simple physical containment of the waste but in some cases also chemically immobilise the waste through incorporation in the phases which form during set[3]. Cements are also relatively cheap, and can be flexible in being easily formulated to cope with aqueous waste streams. Portland cement systems also reduce actinide solubility by the high pH environment.
        Portland cement with addition of blast furnace slag (BFS) or pulverised fuel ash (PFA) has been used extensively[2, 4]. This system combines the robustness of the hydration of Portland cement and takes advantage of the much slower reaction of BFS or PFA to reduce temperature rises so that larger wasteforms can be produced without cracking. It is recognized, however, that the availability of a range of cementitious binders will enhance the ability to effectively treat the wide range of wastes[4]. Therefore alternative cements such as such as calcium aluminate cements, magnesium and calcium phosphate cements, calcium sulphoaluminate cements, alkali activated systems and geopolymers are being investigated as potential candidates[4]. These alternative cement systems may provide some longer term stability advantages with specific wastes known to interact with Portland cement. The work presented here is part of the UK development effort on the use of magnesium phosphate cements to treat ILW.
        Magnesium phosphate cements are an alternative to Portland cement for encapsulation of specific wastes where lower free water content, lower internal pH or the immobilisation of radioactive

107

isotopes as low-solubility phosphates minerals may provide product advantages. To limit the heat generation during setting, normally PFA is added to blends of magnesium oxide and a phosphate source.

Since cement mixtures are almost inevitably multi-component, a large number of variables can be altered in formulating the composition of the paste or slurry. The aim of formulation design is to meet as many of the requirements for the use of the cement as possible. These requirements include criteria for the process by which the cement will be placed as well as requirements on strength development, ultimate strength and durability. For example, for nuclear waste encapsulation, the cement formulation should be highly fluid for a period of up to 2.5 h after initial mixing, the cement should reach a minimum strength of 0.7 MPa at 1 day and thereafter gain and retain sufficient strength to ensure that the containers can be handled for prolonged periods of time.

Developing such formulations can be approached as a process of trial and error, but an understanding of the influence of key parameters on the performance of the mix can greatly reduce the amount of work needed.

The work reported here focusses on establishing the correlation between mix design and strength development in magnesium phosphate cements produced by reacting magnesium oxide (MgO) powder with potassium dihydrogen phosphate (KDP) in mixtures diluted by PFA addition with or without the addition of boric acid ($B(OH)_3$) as a retarder. The overall chemical reaction that causes set and strength development therefore is:

$$KH_2PO_4 + MgO + 5H_2O \leftrightarrow MgKPO_4 \cdot 6H_2O \qquad (1)$$

Competing reactions that could potentially occur are the simple hydration of MgO[1-4] and/or the carbonation to hydrated carbonates of Mg^{5}, or that of MgO and silica in the PFA to yield magnesium-silicate-hydrate gel (M S H gel)[6-9]. For the latter magnesium to silicon ratios in the range of 0.67 to 1 have been observed[10]. Finally depending on the calcium content and availability, the PFA could act as a hydraulic cement by reaction of CaO with SiO_2 and water to give calcium-silicate-hydrate gel[11] or the CaO from the PFA could react with the phosphate present. To date, no evidence for these competing reactions was found and therefore it is assumed that all competing reactions can be ignored for the purposes of predicting the strength. It is recognised, however, that the amorphous nature of the products of some of the competing reactions can make detection difficult, and that refinements could be needed if competing reactions are found to occur significantly.

The approach taken was to investigate the extent of reaction as a function of time for a limited number of mixes and use this information to calculate the reduction in porosity as the cement ages. This allowed determining a master curve for the relation between porosity and strength, which when applied to other mixes should give an indication whether porosity is the main factor driving strength.

EXPERIMENTAL

Material fabrication

The cements were formulated using distilled deionised water, 60 Mesh dead burned MgO (Lehvos UK Limited, UK), potassium dihydrogen phosphate (KH_2PO_4, Prayon UK, UK), pulverised fly ash (Sellafield Ltd. Grade) and boric acid ($B(OH)_3$, Fisher Scientific, UK). Except for the PFA, all materials were >95 % assay. Sample fabrication consisted of addition of the powdered starting materials to the mix water in the order of boric acid, PFA, MgO and finally the phosphate over a period of five minutes while mixing the slurry at low speed (47 rpm) using a planetary mixer. The low speed mixing is continued until 10 minutes have passed, at which time the mixing speed is increased to a high speed (88 rpm) for 10 minutes. A second slow mixing stage is then initiated, the onset of which is arbitrarily defined as the reference time, t = 0, against which all the cement ages are measured. This stage lasts for 150 min, after which point the slurry is poured into a greased mould and the mould face sealed to maintain high humidity conditions by preventing the loss of any mix water by evaporation. The mould is then left at the curing temperature (22 ± 2 °C) until demoulding after 24 hours. The demoulded cement is wrapped in air tight wrapping to continue curing under the same conditions. This method is preferred over curing in a high humidity environment as the uptake of external water is prevented as would be the case when the cement is used for encapsulation of nuclear waste.

There are 5 components in the blends (MgO, KDP, PFA, boric acid and water) and therefore 4 independent variables should suffice to describe the relative composition of the entire blend. A choice was made to express all recipes for 1 mole of potassium dihydrogen phosphate and use molar ratio's of magnesium (M/P) and water (W/P) to phosphate as the first parameters. From equation (1) the theoretical values are M/P=1 and W/P=5. In the literature, the use of excess MgO is quite common[12-14]. However, since excess MgO could potentially lead to delayed hydration of MgO, which is expansive[15, 16], it was desirable to reduce the MgO content. Moreover, in contrast to the use of magnesium phosphate cements in road and runway repair, the nuclear industry requires that initial set does not occur before 4 h, and reducing the MgO content slows the reaction down as the kinetics depend on magnesium ion dissolution. The amount of boric acid, which also slows down reaction by reacting with the surface of the MgO powder and making Mg^{2+} dissolution more difficult[17, 18], is expressed as a mass fraction, λ, of the MgO, and the PFA content is defined by also reporting a water to solids mass ratio (w/s). The latter was chosen over a direct reporting of the amount of PFA as it is indicative of the fluidity, which for waste encapsulation needs to be high. The compositions of the formulations investigated are shown in Table 1. For ease of interpretation, the actual recipe in gram of each material was also included in the table.

Table 1 Formulations of the different mixes. For the first 10 mixes, parameters in bold indicate variations on the reference mix 21. Mix 74, 75 and 76 were formulated afterwards using previous results.

Mix	M/P	W/P	λ	w/s	KDP	MgO	H_2O	$B(OH)_3$	PFA
					g	g	g	g	g
21	1.5	5.89	0.09	0.30	136.1	60.5	106.1	5.4	151.3
25	1.5	**4.15**	0.09	0.22	136.1	60.5	74.7	5.4	135.6
26	1.5	**5.02**	0.09	0.26	136.1	60.5	90.4	5.4	143.5
27	1.5	5.89	**0.00**	0.31	136.1	60.5	106.1	0.0	151.3
28	1.5	5.89	**0.18**	0.30	136.1	60.5	106.1	10.9	151.3
29	**1.0**	5.29	0.09	0.30	136.1	40.3	95.3	3.6	135.8
31	**1.25**	5.59	0.09	0.30	136.1	50.4	100.7	4.5	143.6
210	2.0	6.50	0.09	0.30	136.1	80.6	117.0	7.3	166.9
212	1.5	5.89	0.09	**0.38**	136.1	60.5	106.1	5.4	75.7
213	1.5	5.89	0.09	**0.25**	136.1	60.5	106.1	5.4	227.0
74	1.75	5.05	0.18	0.23	136.1	70.5	90.9	12.7	178.5
75	1.75	5.05	0.18	**0.25**	136.1	70.5	90.9	12.7	**148.8**
76	1.75	**5.51**	0.18	**0.25**	136.1	70.5	99.9	12.7	183.5

Mix "21" was close to a mixture used in previous investigations[19, 20] and was used as a starting point. A systematic range of variations was tested: altering the W/P ratio at fixed PFA mass fraction relative to the cement (mixes 25 & 26), altering the boric acid content (mixes 27 & 28), altering the M/P ratio (mixes 29 & 31) and altering the w/s ratio by changing the PFA content (mixes 212 & 213). Mixes 74-76 were formulated for further work on the basis of the initial study. Both mixes 75 and 76 have slightly higher water to solids ratios than mix 74. For mix 75 the water to solids ratio was increased by reducing the PFA content, whereas for mix 76 this was achieved by increasing the W/P ratio.

Characterisation

For the initial mixes (21-213), the majority of the work was based on determining the processing characteristics, which have been reported elsewhere[20], x-ray diffraction for identification of crystalline phases present and compressive strength measurements. The latter were normally carried out on 40 mm × 40 mm × 40 mm cubes on a Zwick 1474 100kN universal test frame using a position controlled loading rate of 10 mm min^{-1} at 1, 7, 28, 90 and 180 days. For high strength samples (>60 MPa), smaller samples (20 mm × 20 mm × 20 mm) had to be used to remain within the range of the load cell.

For mixes 74-76, this was supplemented by combined thermogravimetric and differential thermal analysis (TGA/DTA, STA 449 F1 Jupiter, Netzsch, Germany). The aim of the former was to determine the extent of reaction between the phosphate and the magnesium oxide from the weight loss when the magnesium phosphate crystals de-hydrate. However, since the hydrated magnesium potassium phosphate decomposes and loses its water near 100 °C, free water had to be removed in a different way. Previous work had shown that crystalline water is not lost below 50 °C[20], but it was expected that heating could accelerate the reactions. Therefore the drying procedure consisted of grinding the cement and quenching in acetone followed by drying and storage in vacuum (<100 Pa). Such a method is commonly used to stop hydration of Portland cement and as will be discussed below seems useful for MKP cements too.

RESULTS

The combined TGA/DTA data for mix 74 after 1 day and 7 days are shown in Figure 1. Up to 500 °C, there is only one weight loss step (100 °C – 200 °C) due to the dehydration of the magnesium potassium phosphate hexa-hydrate (MKP) as 6 moles of water are lost per mole of MKP. The small shoulder in DTA peak near 100 °C in the 1 day data indicates that perhaps not all free water was removed by the drying procedure, but its absence in the 7 day result and all older mixes suggest that this was not a problem for later ages.

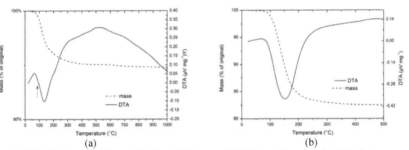

Figure 1. TGA/DTA results for mix 74 after (a) 1 day and (b) 7 days. The arrow in (a) indicates a small shoulder on the endothermic peak associated with water loss from the sample.

Since progression of the reaction creates 1 mole of water and binds 5 moles of water into the crystals, the extent of reaction, α:

$$\alpha = \frac{n_{MKP}}{n_{MKP,max}} \qquad (2)$$

where n_{MKP} is the number of moles of MKP formed and $n_{MKP,max}$ is the maximum number of moles of MKP that can theoretically form can be calculated from the weight loss near 100 °C. Figure 2 shows that initially the reaction is quite fast but it slows down considerably. The curve shown in Figure 2 illustrates that the progression of the reaction can be fitted reasonably with:

$$\alpha = 1 - \exp\left(-\left(\frac{t}{a}\right)^{n}\right) \qquad (3)$$

with t the time in days, a = 3.85 days and n = 0.282.

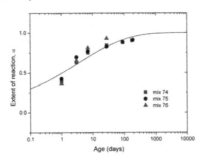

Figure 2. The extent of reaction, α, as a function of time for mixes 74-76 (symbols) and the fit to eq.3 (solid line).

Examples of the stress strain curves obtained during compressive testing of mix 75 at different ages are shown in Figure 3a. After limited curing times (≤ 7 days), the material is not strongly cohesive and deforms like a soil. For longer curing times (> 7 days), the stiffness is much higher and the samples show brittle failure. This transition in behaviour was seen for all the mixes, albeit not necessarily at the exact same ages.

The strength results of all mixes have been plotted as a function of curing time in Figure 3b. Despite a fairly limited variation in the formulations, the strengths achieved vary by almost an order of magnitude from 70 MPa to below 10 MPa after 180 days.

DISCUSSION

It is well known that porosity plays a key role in determining the mechanical properties of ceramics[21-24] and cements[25-29]. Therefore if the extremely varied strength results shown in Figure 3b are to be understood, a normalisation for porosity must be made. This can be achieved conveniently by calculation provided the degree of reaction is known to allow accounting for the change in the nature of the solids and their density during the chemical reaction. Hence for mixes 74-76 where the extent of reaction as a function of time was measured using TGA, the porosity in these mixes as a function of time can be calculated. The calculation requires the density and molar mass of the different phases to be known, see Table 2, and assumes that the change in the overall volume of the cement is small enough to be negligible in estimates of the density. This assumption was in fact confirmed by measurements of the dimensions of the samples during ageing.

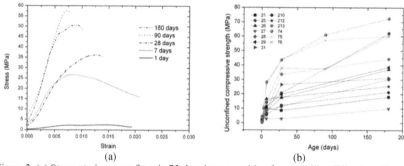

Figure 3. (a) Stress-strain curves for mix 75 showing a transition from soil-like sliding to stiffer and more brittle behaviour with age and (b) Strength versus age for all mixes. The lines in (b) were drawn to enable a better judgement of the data points in the same series and have no further meaning.

It is then straightforward to calculate the total volume from the formulation, and to proceed to calculate the volume fraction solids using eq.3 to calculate how much of the KDP, MgO and water have reacted to form MKP. Figure 4 shows a plot of the strength of these mixes as a function of the calculated volume fraction solids. Since the volume fraction of solids can also be regarded as the relative density, the latter term will be used here.

Table 2 Molar mass and density of the different minerals and chemicals in the cement formulations

Material	Molar mass (g mol^{-1})	Density (g cm^{-3})
MgO	40.30	3.58
KDP	136.09	2.34
H_2O	18.02	1.00
MKP-6H_2O	266.47	1.86
B(OH)$_3$	61.83	1.435
Mg(OH)$_2$	58.32	2.44
PFA	n/a	2.10

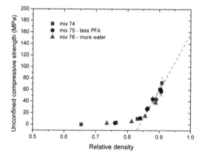

Figure 4. Strength versus relative density.

Interestingly, the calculation shows that despite rapid initial reaction – 40% of the MKP forms during the first day - the strength increases are limited until the relative density has increased by 0.2. Thereafter, there is a strong linear increase in the strength with relative density. The latter is consistent with the observations of Lam, Lange and Evans[30], who found that during sintering of ceramic bodies the mechanical properties scale approximately linearly with $1-P/P_0$, where P is the volume fraction of pores and P_0 is the volume fraction of pores in the initial body. This is reasonable because in a ceramic body, even before sintering, the particles are in contact. When densification starts, necks form between particles and on average the contact area increases proportional to the relative densification. More generally, Phani and Niyogi[31] proposed that for example the elastic modulus, E, should evolve with a reduction in porosity, P, according to expressions of the form :

$$E = E_0 \cdot \left(1 - \frac{P}{P_{crit}}\right)^n \qquad (4)$$

Where P_{crit} is a critical volume fraction where the elastic modulus goes to zero and E_0 is the elastic modulus of the dense body. P_{crit} and the exponent n are geometrical factors, which depend on the initial packing and the distribution of particle sizes in the mix. Similar relationships have been used to describe the variation of properties of cements including compressive strength, see e.g.[28, 32]. In contrast to sintering of a pressed ceramic body, the MKP cements investigated here are formulated to give a fluid slurry and therefore the particles are not in contact. Since these formulations also do not bleed, the initial reaction product that forms cannot bind particles together. Therefore, in the initial stage, the reaction products merely gel the liquid by making fluid flow more difficult. Strengthening only becomes substantial when a sufficient amount of solid has formed and the particle-particle contacts start to thicken. This is represented schematically in Figure 5. The fact that there are two stages in strengthening is also consistent with the transition in the stress-strain curves from soil-like behaviour in the gelation stage to linear elastic and brittle behaviour once substantial inter-particle links have formed.

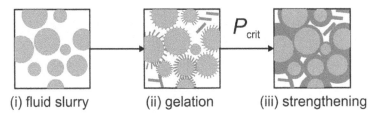

(i) fluid slurry   (ii) gelation   (iii) strengthening

Figure 5. Schematic showing how when starting from a fluid slurry in which particles are not touching (i), the first solid to form reduces the free space for water to flow in making the body gel but does not yet lead to strong inter-particle contacts. Therefore substantial strengthening only occurs when a critical porosity level, P_{crit}, is exceeded and significant particle-particle bonding develops.

Extrapolation of the rapid strengthening data in Figure 4 suggests that the critical volume fraction porosity in these systems is about 0.163, or a relative density of 0.837. This is rather high compared to 0.66 for random packing of monosized spheres, but not impossible when considering that there is a wide range of particle sizes in the mix and that even simple random packing of small sized particles in between random packed bigger particles leads to an expected packing fraction of 0.88.

Moreover, this high packing density is reached by in-situ formation of material and not by actual packing of particles.

It is therefore appropriate to divide the strength evolution into two regimes: (i) a gelation region followed by (ii) a region where strengthening is proportional to the extent of densification. For the former a simple power law was fitted to the experimental data, whereas for the latter, an expression of the form proposed by Lam, Lange and Evans[30] was used:

Gelation $(P > P_{crit})$:

$$\sigma = A(1 - P)^n \tag{5}$$

Strengthening $(P \leq P_{crit})$:

$$\sigma = \sigma_0 \cdot \left(1 - \frac{P}{P_{crit}}\right) \tag{6}$$

The values obtained by fitting to the data where A=69.4 MPa, n = 11.4, Pcrit = 0.163 and σ_0 129.6 MPa.

Eq.5 and eq.6 in combination with eq.3 are sufficient to make estimates for the strength of any mix at any age directly from its formulation. Since the expressions were derived for the average behaviour of mix 74 to 76, their strength evolution with time should be predicted quite well. Figure 6a illustrates that this is the case by comparing the experimental data with the predictions for mixes 74-76.

Although a good fit was to be expected, the modelled curve, with its distinct two stages, aids in understanding why initial strengthening is so slow: in these strongly retarded systems the gelation stage extends to several days. Such a slow development of strength is markedly different from the rapid set normally seen in magnesium phosphate cements[18, 33]. That such long gelation times and hence limited initial strengthening are quite real is borne out by mixes 29 & 31, which have lower M/P ratios, and where the gelation stage exceeds 10 days before marked strengthening occurs, see Figure 6b.

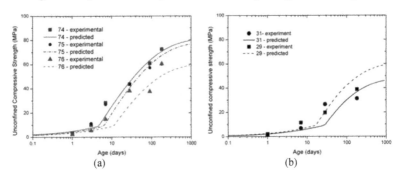

Figure 6. Comparison of predictions of strengthening with experimental observations: (a) mix 74, 75 & 76, (b) mix 29 and 31 with low M/P

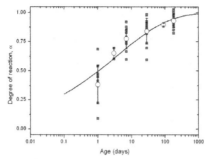

Figure 7. Degree of reaction for the individual mixes (■) and average degree of reaction for all mixes (○± standard deviation) as a function of time obtained by calculation: α was varied until the predicted strength was equal to the experimental strength. Also shown is the line representing eq. 3, i.e. the best fit to the measured α for mixes 74-76.

Another observation is that the model calibrated with only limited data from 3 mixes, predicts other mixes reasonably. The latter can be illustrated also by reversing the analysis: instead of assuming the degree of reaction, α, and calculating the strength, it is also possible to use the strength to estimate the degree of reaction by varying α until the calculated strength using eq. 5 and 6 is equal to the measured strength. As shown in Figure 7, on average the estimates for the degree of reaction obtained in this way agree rather well with the line representing the fit to the measured degree of reactions.

This indicates that the wide variation in strength observed in Figure 3b is in fact a simple representation of the effect of formulation on the relative density of the final product, i.e. strength is determined much more strongly by porosity than by the exact make-up of the solids in the mix.

CONCLUSIONS

Measurements of the strength development and of the degree of reaction of three mixes were used to develop an empirical description based on theoretical relations of the strength development in highly fluid and strongly retarded magnesium phosphate cement formulations.

It was found that due to the highly fluid nature of the mixes, strengthening occurs in two stages: material formed during the first gelation stage, make fluid flow more difficult but give limited cohesion. The second stage only starts when sufficient solid has formed to bind the particles together.

The empiric model calibrated with only 3 mixes was found to describe the strength development in a wider range of highly fluid mixes quite well. Therefore the main factor determining the strength of these formulations is the residual porosity, which can be predicted easily.

ACKNOWLEDGEMENTS

William Montague thanks the Engineering and Physical Sciences Research Council of the UK and the National Nuclear Laboratory for funding him through a CASE award.

REFERENCES

1.   M.I. Ojovan and W.E. Lee, An introduction to nuclear waste immobilisation, London: Elsevier. 315 (2005).
2.   R.J. Caldwell, S. Rawlinson, E.J. Butcher, and I.H. Godfrey. Characterisation of full-scale historic inactive cement-based intermediate level wasteforms. in *Stabilisation/Solidification Treatment and Remediation*. Cambridge: A.A. Balkema Publishers (2005).

3.   C.M. Jantzen, F.P. Glasser, and E.E. Lachowski, Radioactive waste-Portland cement systems: I, Radionuclide distribution. *Journal of the American Ceramic Society*, **67**(10): p. 668-673 (1984).
4.   N.B. Milestone, Reactions in cement encapsulated nuclear wastes: need for toolbox of different cement types. *Advances in Applied Ceramics*, **105**(1): p. 13-20 (2006).
5.   L.J. Vandeperre and A. Al-Tabbaa, Accelerated carbonation of reactive MgO cements. *Advances in Cement Research*, **19**(2): p. 67-79 (2007).
6.   T. Zhang, C.R. Cheeseman, and L.J. Vandeperre, Development of low pH cement systems forming magnesium silicate hydrate (M-S-H). *Cement and Concrete Research*, **41**(4): p. 439-442 (2011).
7.   T.T. Zhang, L. Vandeperre, and C.R. Cheeseman, Bottom-up design of a cement for nuclear waste encapsulation. *Ceramic Engineering and Science Proceedings*, **32**(9): p. TBC (2011).
8.   T.T. Zhang, C.R. Cheeseman, and L.J. Vandeperre, Characterisation of corrosion of nuclear metal wastes encapsulated in magnesium silicate hydrate (MSH) cement. *Ceram. Eng. Sci. Proc.*, **33**(9): p. in press (2012).
9.   D.R.M. Brew and F.P. Glasser, The magnesia-silica gel phase in slag cements: alkali (K, Cs) sorption potential of synthetic gels. *Cement and Concrete Research*, **35**: p. 77-83 (2005).
10.   D.R.M. Brew and F.P. Glasser, Synthesis and characterisation of magnesium silicate hydrate gels. *Cement and Concrete Research*, **35**: p. 85-98 (2005).
11.   S. Mindess, J.F. Young, and D. Darwin, Concrete. second ed, Upper Saddle River: Prentice Hall. 644 (2003).
12.   Z. Ding, B. Dong, F. Xing, N. Han, and Z. Li, Cementing mechanism of potassium phosphate based magnesium phosphate cement. *Ceramics international*, **38**: p. 6281-6288 (2012).
13.   J. Formosa, J.M. Chimenos, A.M. Lacasta, and M. Niubo, Interaction between low-grade magnesium oxide and boric acid in chemically bonded phosphate ceramics formulation. *Ceramics international*, **38**: p. 2483-2493 (2012).
14.   F. Qiao, C.K. Chau, and Z. Li, Calorimetric study of magnesium potassium phosphate cement. *Materials And Structures*, **45**: p. 447-456 (2012).
15.   L.S. Wells, W.F. Clarke, and E.M. Levin, Expansive characteristics of hydrated limes and the development of an autoclave test for soundness. *Journal of Research of the National Bureau of Standards*, **41**: p. 179-204 (1948).
16.   L. Vandeperre, M. Liska, and A. Al-Tabbaa. Reactive MgO cements : Properties and Applications. in *International conference on sustainable construction materials and technologies*. Coventry: Taylor and Francis (2007).
17.   Q. Yang and X. Wu, Factors influencing properties of phosphate cement-based binder for rapid repair of concrete. *Cement and Concrete Research*, **29**: p. 389-396 (1999).
18.   S.R. Iyengar and A. Al-Tabbaa, Developmental study of a low pH magnesium phosphate cement for environmental applications. *Environmental Technology*, **28**: p. 1387-1401 (2007).
19.   A. Covill, N.C. Hyatt, J. Hill, and N.C. Collier, Development of magnesium phosphate cements for encapsulation of radioactive waste. *advances in Applied Ceramics*, **0**(0): p. 0
20.   W. Montague, L. Vandeperre, and M. Hayes, Processing characteristics and strength of magnesium phosphate cement formulations compatible with UK nuclear waste treatment plants, in *Scientific Basis for Nuclear Waste Management XXXV* (2012).
21.   R.W. Rice, Porosity of ceramics. Materials Engineering, New York: Marcel Dekker Inc. 539 (1998).
22.   L.J. Vandeperre, J. Wang, and W.J. Clegg, Effects of porosity on the measured fracture energy of brittle materials. *Philosophical magazine*, **84**(34): p. 3689-3704 (2004).
23.   S. Raghavan, H. Wang, R.B. Dinwiddie, W.D. Porter, and M.J. Mayo, The effect of Grain Size, porosity and yttria content on the thermal conductivity of nanocrystalline zirconia. *Scripta Materialia*, **39**(8): p. 1119-25 (1998).
24.   R.L. Coble and W.D. Kingerey, Effect of Porosity on Thermal Stress Fracture. *Journal of the American Ceramic Society*, **38**(1): p. 33-37 (1955).
25.   T. Matusinovic, J. Sipusic, and N. Vrbos, Porosity-strength relation in calcium aluminate cement pastes. *Cement and Concrete Research*, **33**(11): p. 1801-1806 (2003).

26. K. Kendall, A.J. Howard, J.D. Birchall, P.L. Pratt, B.A. Proctor, and S.A. Jefferis, The relation between porosity, microstructure and strength, and the approach to advanced cement-based materials. *Phil. Trans. R. Soc. Lond.*, **A310**(1511): p. 139-153 (1983).

27. M. Rößler and I. Odler, Investigations on the relationship between porosity, structure and strength of hydrated Portland cement pastes I. Effect of porosity. *Cement and Concrete Research*, **15**(2): p. 320-330 (1985).

28. Y.-X. Li, Y.-M. Chen, J.-X. Wei, X.-Y. He, H.-T. Zhang, and W.S. Zhang, A study on the relationship between porosity of the cement paste with mineral additives and compressive strength of mortar based on this paste. *Cement and Concrete Research*, (2006).

29. L.J. Vandeperre, M. Liska, and A. Al-Tabbaa, Hydration and Mechanical Properties of Magnesia, Pulverized Fuel Ash, and Portland Cement Blends. *Journal of Materials in Civil Engineering*, **20**: p. 375-383 (2008).

30. D.C.C. Lam, F.F. Lange, and A.G. Evans, Mechanical Properties of Partially Dense Alumina Produced from Powder Compacts. *Journal of the American Ceramic Society*, **77**(8): p. 2113-17 (1994).

31. K.K. Phani and S.K. Niyogi, Elastic modulus-porosity relation in polycrystalline rare-earth oxides. *Journal of the American Ceramic Society*, **70**(12): p. C362-C366 (1987).

32. T.C. Hansen, Influence of Aggregates and Voids on Modulus of Elasticity of Concrete, Cement Mortar, and Cement Paste. *Journal of the American Concrete Institute*: p. 193-216 (1965).

33. A.D. Wilson and J.W. Nicholson, Acid-base Cements - Their biomedical and industrial applications. Chemistry of Solid State Materials, ed. A.R. West and H. Baxter, Cambridge: Cambridge University Press (1993).

TEST METHODS FOR HOOP TENSILE STRENGTH OF CERAMIC COMPOSITE TUBES FOR LIGHT WATER NUCLEAR REACTOR APPLICATIONS

Michael G. Jenkins, Bothell Engineering & Science Technologies, Fresno, CA, USA,
jenkinsmg@bothellest.com
Jonathan A. Salem, NASA Glenn Research Center, Cleveland, OH, USA,
jonathan.a.salem@nasa.gov

ABSTRACT
The US DOE plans to replace conventional zirconium-alloy fuel rod tubes in light water reactors (LWR) with those consisting of ceramic matrix composites (CMC) thereby enhancing fuel performance and accident tolerance of LWRs. Silicon carbide fiber-reinforced silicon carbide-matrix (SiC/SiC) composites demonstrate tolerance to the irradiation and chemical environments of LWRs. Loss of gas tightness and mechanical integrity due to the build-up of internal gas pressure and the swelling of fuel pellets are among the anticipated failure modes for the LWR fuel cladding. Therefore, rigorous determination of the hoop tensile (or equivalent) strength properties is critically important for evaluation of SiC/SiC CMC fuel claddings. Because there are no commonly-accepted design methodologies for advanced composite tubular components, there are limited mechanical test standards for any properties of tubular ceramic composite components. Therefore, some current and proposed test methods for measuring tensile hoop strength of composite tubes are presented, discussed, and compared for application to CMCs. Proposed standard test methods are presented in terms of the following experimental issues -- test specimen geometries/preparation, test fixtures, test equipment, interferences, testing modes/procedures, data collection, calculations, reporting requirements, and precision/bias.

KEYWORDS – hoop tensile strength, tubes, ceramic matrix composite, silicon carbide composite, nuclear fission.

INTRODUCTION

The US Department of Energy (US DOE) is currently exploring replacing conventional zirconium-alloy fuel rod tubes in light water reactors (LWR) with fuel rods consisting partly or entirely of ceramic matrix composites (CMC) thereby benefiting LWRs by enhancing fuel performance and accident tolerance. The specific CMC of interest for this application is silicon carbide continuous fiber-reinforced silicon carbide-matrix (SiC/SiC) composite because of the demonstrated tolerance of SiC/SiC CMC for the irradiation and chemical environment of LWRs. In particular, high strength at high temperatures and low chemical activity, including no exothermic reaction with water as zirconium demonstrates at elevated temperatures, were the primary reasons to select SiC/SiC CMCs for further LWR development. Additionally, the high temperature properties of SiC/SiC CMC imply that the fuel system can retain its geometry and fuel protective function even during an accident. Removal of the exothermic zirconium and water reaction also increases the temperature at which the fuel can operate. Eliminating the generation of free hydrogen would also lower the type of risks created during an accident scenario [1,2].

However, loss of gas tightness and mechanical integrity due to the build-up of internal gas pressure and the swelling of fuel pellets are among the anticipated failure modes for the LWR fuel cladding. Therefore, rigorous determination of the hoop tensile (or equivalent) strength properties is critically important upon evaluation of the SiC/SiC CMC fuel cladding. These CMCs consist of high-strength silicon carbide fibers in a high-temperature silicon carbide matrix. Such a composite structure provides high strength and high fracture resistance at elevated temperatures, in addition to their potentially higher resistance to neutron radiation [5,6] than conventional material.

Zircaloy-4   SiCf/Sic

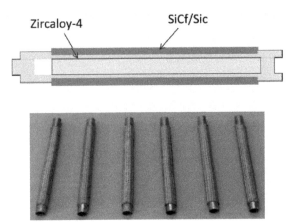

Figure 1 SiC/SiC CMC cladding for LWR fuel rods (from Ref 1)

The ceramic reinforcement tows have high filament counts (500-2000) and are woven with large units cells, several millimeters in size. In a tube configuration, the composites can have a 1-D filament wound, 2-D laminate, or 3-D (weave or braid) construction depending on what tensile, shear, and hoop stresses are considered. The fiber architecture in the tubes can be geometrically tailored for highly anisotropic or uniform isotropic mechanical and thermal properties.[2, 3]

Tubular geometries for nuclear applications present challenges for both "makers" and "lookers" of SiC/SiC CMCs. For "makers": how to make seamless tubes with multiple direction architectures; how to ensure integrity in the radial direction; and how to create uniform wall thickness and uniform/nonporous matrices. For "lookers": how to build on decades of experience with consensus standards and data bases for "flat" material forms; how to interpret information of tests of test specimen in component form; and how to adapt expertise at room temperature in ambient environments to conditions at high temperature in specific extreme-use environments.

However, not only are there are no commonly-accepted design methodologies for tubular components comprised of advanced composites, at this time there are almost no mechanical test standards for any of these properties of tubular geometry ceramic composite components. In particular, for CMC tubes there is one standard for axial tensile strength that was currently being approved and published by ASTM as C1624-12 "Standard Test Method for Monotonic Tensile Behavior of Continuous Fiber-Reinforced Advanced Ceramic Tubular Test Specimens at Ambient Temperature."

Use of new CMC materials in LWR applications requires mechanical test standards to support not only material development and property databases, but design codes and component specification documents, as well as Nuclear Regulatory Commission (NRC) regulations on nuclear design approval, certification, and licensing[4, 5]. In particular, the mechanical test standards for nuclear grade CMCs are necessary to provide for accurate and reliable data, based on well-defined test methods, detailed specimen preparation, comprehensive reporting requirements, and commonly-accepted terminology. The development and component design process using CMCs in LWR applications will be hampered and delayed, if appropriate CMC mechanical test standards are not available in a timely manner.

Fortunately the LWR applications of SiC/SiC CMCs builds on experience allowing nuclear applications to advance an existing mature specialized technology. For example, SiC/SiC CMC materials and structure technology were funded by the aerospace and defense industries/agencies. In addition, current evaluations and applications of SiC/SiC CMCs in fusion reactors (first wall) and tristructural-isotropic (TRISO) fuel forms that have established properties under extended neutron irradiation and at high temperatures as well as very hot steam environment. Growing, credible data bases for SiC/SiC CMCs now exist because of the evolution of consensus test methods and design codes. Finally, maturation of volume-scale manufacturing capability for all types of CMCs including SiC/SiC CMC adds to availability and understanding of these material.

Such professional organizations as ASME and ASTM have taken the lead in developing the codes, specifications, and test standards for CMCs in nuclear applications. ASTM Committee C28 on Advanced Ceramics has a particular focus on mechanical test standards for CMCs. Specifically, ASTM Subcommittee C28.07 has published eleven standards for CMCs (e.g., tensile, flexure, shear, compression, creep, fatigue, etc.

Mechanical testing of composite tube geometries is distinctly different from testing flat plates because of the differences in fiber architecture (weaves, braids, filament wound), stress conditions (hoop, torsion, and flexure stresses), gripping, bending stresses, gage section definition, and scaling issues.[5] Because there are no commonly-accepted design methodologies for advanced composite tubular components, there are almost no mechanical test standards for any properties of tubular ceramic composite components.

Therefore, in this paper, some current and proposed test methods for measuring tensile hoop strength of composite tubes are presented, discussed, and compared for application to CMCs. Two proposed test methods are presented in terms of the following experimental issues -- test specimen geometries/preparation, test fixtures, test equipment, interferences, testing modes/procedures, data collection, calculations, reporting requirements, and precision/bias.

HOOP TENSILE STRENGTH TEST OF TUBES

A review of the literature for experimental and analytical methods applied to assessing behavior of tubes subjected to hoop tensile stress resulted in the following categories.

1) Mechanical loading methods applied to short sections of tubes
2) Viscoelastic loading methods applied to short and/or long sections of tubes
3) Pressure loading methods applied to long sections of tubes

Aspects of each category are discussed and illustrated in the following sub sections.

Mechanical loading methods applied to short sections of tubes - Longitudinally "short" sections of tubes are loaded transversely through split disk loading fixtures as illustrated in Fig. 2. This method has been standardized in ASTM D2290[6] and previous work has shown that this developed test, compared to the quick burst test, induces a hoop (circumferential) stress, which is similar to the stress induced by internal pressure[7]. However, failures from these types of tests tend to initiate from the inner radius and edges of the short sections that may not be representative of failures of actual long tubes. Some pros and cons for this type of test are listed in Table 1.

Table 1 Some Pros and Cons for Mechanical Loading Methods Applied to Short Sections of Tubes

Pros	Cons
- Simple fixtures	- Samples only part of tube
- Uses small sections of tube	- Edge effects
- Uses existing test machines	- Does not represent internal pressure loading
- Simple equations	- Limited to proof testing

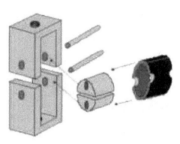

Figure 2 Illustration of split disk mechanical loading method applied to short section tube (from Ref 7)

Viscoelastic loading methods applied to short and/or long sections of tubes - This is a simple method to generate high-pressure conditions for pressure testing small-diameter tubes using the radial expansion of an axially-compressed viscoelastic insert. At room temperature, a piston is used to compress an elastomeric cylinder or plug inside a tube of material. The resulting Poisson's expansion generates radial pressure along the inner wall of the tube sample. One of the attractive attributes of this test is that once the sample fails, the elastomer easily compresses and quickly lowers the stress and pressure in the system[8-10]. Additionally, there are no high-pressure gases or fluids to contain. Also, the use of a solid material to generate the internal pressure removes the need for high-pressure seals. This method has been extended to high temperatures by using a glass insert material that behaves viscoelastically above its glass transition temperature[10]. Some pros and cons for this type of test are listed in Table 2.

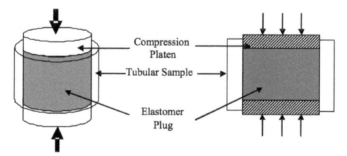

Figure 3 Illustration of elastomeric insert applied (viscoelastic loading) applied to short section of tube (from Ref 8)

Table 2 Some Pros and Cons for Viscoelastic Loading Methods Applied to Short and/or Long Sections of Tubes

Pros	Cons
- Simple fixtures	- May be complex stress states
- Uses short or long sections of tube	- Friction effects for rough surfaces
- Uses existing test machines	- May be force/pressure limited
- Has been extended to high temp	- May be limited to material selection

Pressure loading methods applied to short or long sections of tubes - This is a conceptually simple but experimentally-challenging method for pressure testing tubes using internal pressurization. Pressure can be applied using either gas or liquid. In addition, the pressurized medium can be applied directly to the tube or through an internal bladder. Finally, end caps can be attached to the side walls of the tube or to each other. Experimental conditions that have produced consistent hoop stresses in composite tubes include the following: pressurized liquid to minimize explosive failures upon rupture; internal bladders to prevent leakage through the composite tube walls; and end caps attached to each other and not to the tube walls to eliminate axial stresses [11-23]. ASTM D1599 uses this method for "plastic" pipes[23]. Using tubes with proper length to diameter ratios, end effects are eliminated and the hoop tensile stresses and strains are uniform in the gage section, thereby resulting in measure of the material properties of the tubular material during testing. With proper configuration, these test methods can be extended to elevated temperature [12]. Some pros and cons for this type of test are listed in Table 3.

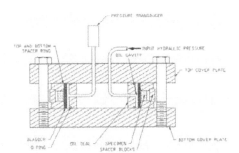

Figure 4 Illustration of internally pressured tube with closed ends resulting in hoop and axial stresses (from Ref 12)

Various studies of homogenous, isotropic, linear elastic materials such as low carbon steel test materials using methods from each of the three techniques have indicated that the elastic response and the strength behavior determined from hoop stresses and strains are comparable regardless of test method. However, the complexity of the non homogeneous, anisotropic, nonlinear elastic behaviour of fibre-reinforced composites has led to differences in results from the different methods. It should be noted that the potential biaxiality and even triaxility of the stress states may limit interpretation of these tests methods for "pure" hoop behavior. Therefore, the most appropriate test methods for ceramic composites at room temperature are those that result in hoop stresses that arise from internal pressurization. The evolving standardized test method for hoop tensile strength of ceramic composites described in the following sections reflects this conclusion.

Table 3 Some Pros and Cons for Pressure Loading Methods Applied to Short or Long Sections of Tubes

Pros	Cons
- Real internal pressure loading	- May require internal bladders for non impervious materials
- Uses short or long sections of tube	- Stress state may be biaxial
- Simple equations	- May require special equipment
- Related directly to applications	- May be high-temp problematic

SCOPE AND APPLICATION

A working group of ASTM Subcommittee C28.07 on CMC tube testing is developing two draft test methods: 1) "Hoop Tensile Behavior of Continuous Fiber-Reinforced Advanced Ceramic Composite Tubular Test Specimens at Ambient Temperature" and 2) "Hoop Tensile Strength of Continuous Fiber-Reinforced Advanced Ceramic Composite Tubular Test Specimens at Ambient Temperature Using Elastomeric Inserts."

The first proposed test method has wide applicability with the test results intended for design data generation as well as model verification. The second proposed test method has limited applicability with the test results intended for material down selection / screening.

In the two test methods, a ceramic composite tube/cylinder or tube/cylinder section with a defined gage section and a known wall thickness is selected to be the test specimen. The test specimen is inserted into the appropriate test fixture assembly is subject to one of the following monotonic loadings depending on the test method:

1) Direct internal hydrostatic pressure produced from hydraulic fluid

or

2) Indirect pressure produced by axial loading of an elastomeric insert

Either pressure or axial load is recorded along with hoop displacement/strain in the gage section. Results include hoop tensile stress/strain, ultimate hoop tensile strength, fracture hoop tensile strength and proportional limit hoop tensile stress along with corresponding strain, and elastic constants. The test method addresses test equipment, interferences, gripping and coupling methods, testing modes and procedures, tubular test specimen geometries, test specimen preparation and conditioning, data collection, calculations, reporting requirements and precision/bias. The methods are applicable to a wide range of CMC tubes with 1-D filament, 2-D laminate, and 3-D weave and braid architectures. In addition, the test methods reference test procedures and fixturing from research work done on CMC tubes as well as test procedures and research on PMC tubes.

EXPERIMENTAL FACTORS

CMCs generally exhibit "graceful" failure from a cumulative damage process, unlike monolithic advanced ceramics that fracture catastrophically from a single dominant flaw. The testing of CMC (both flats and tubes) has a range of different material and experimental factors that interact and must be controlled and managed (See Fig. 5). These factors must be managed and understood to produce consistent, representative failures in the gage section of test specimens. Tubular test specimens with cylindrical geometries provide particular challenges in the areas of gage section geometry, loading and bonding failures, extraneous "parasitic" stresses (including biaxial and triaxial stresses), and out-of-gage failures.

- Material Variability, including Anisotropy, Porosity, and Surface Condition
- Test Specimen Size, Fiber Architecture, and Gage Section Geometry Effects
- Out-Of-Gage Failures and Extraneous Stresses
- Slow Crack Growth, Strain Rate Effects, and Test Environment
- Accurate Strain/Elongation Measurement

Figure 5 Range of "interferences" in test CMC materials

TEST SPECIMEN GEOMETRIES

Test Specimen Size -- CMC tubes are fabricated in a wide range of geometries and sizes, across a spectrum of fiber-matrix-architecture combinations. It is not practical to define a single test specimen geometry that is universally applicable. The selection and definition of a test specimen geometry depends on the purpose of the testing effort. With that consideration, the test method is generally applicable to tubes with outer diameters (D_o) of 10 to 150 mm and wall thicknesses (t) of 1 to 25 mm, where the ratio of the outer diameter to wall thickness is commonly $D_o/t = 5$ to 30. Tube sections may vary depending on the type of test (e.g., 25 mm to 1000 mm). In many cases, the wall thickness is defined by the number of plies and fiber-reinforcement architecture, particularly for woven and braided configurations.

Gage Section Geometry -- Tubular test specimens are classified into two groups – straight-sided and contoured gage-section, as shown in Figures 6 and 7. Contoured gage-section test specimens are distinctive in having gage sections with thinner wall thicknesses than in the grip sections.

Although straight-sided test specimens are easier to fabricate and are commonly used, tubular test specimens with contoured gage sections are preferred to promote failures in the uniformly-stressed gage section.

Experience has shown that successful tests can be maximized by using consistent ranges of relative gage section dimensions, as follows:

$$2 < L_O / D_o < 3 \qquad and \qquad 15 < L_O / t < 30 \qquad (1)$$

where L_O is the defined gage length, D_o is the outer diameter in the gage section, and t is the wall thickness in the gage section of the tube. Deviations from the recommended geometries may be necessary depending upon the particular composite tube geometry being evaluated.

Fig. 6 – Schematic of straight-sided tubular test specimen (from Ref 25)

Fig. 7 – Schematic of contoured gage section tubular test specimen (from Ref 25)

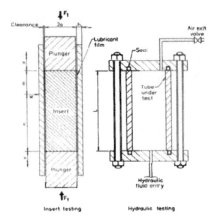

Figure 8 Illustrations of test setups for i) Insert testing and ii) Hydraulic testing of tubes (from Ref. 9)

TEST EQUIPMENT AND PROCEDURES

Test setup, Force and Strain Measurement, Data Acquisition -- The test method can use a standard load frame with a hydraulic or screw drive loading mechanism and standard force transducers for controlled axial loading for the elastomer insert test method. Guidance is given regarding type, composition and properties of the elastomeric insert material. However, a source of controlled-pressure hydraulic fluid is necessary for the internal pressure, hydraulic test method. Primary strain measurement can be measured by strain gages and/or string extensometers in the "gage section." If required, an environmental test chamber may be used to control humidity and ambient temperature. Data collection should be done with a minimum of 50-Hz response and an accuracy of ±0.1 % for all data.

Test Procedures -- Generally, a test mode is used to avoid "run-away" tests that sometimes occur in force-control tests. Test mode rates are chosen so as to produce test specimen failures in 5-50 s. Failure within one minute or less should be sufficient to minimize slow-crack growth (SCG) effects. If slow crack growth is observed (e.g. under slow test mode rates), subsequent tests can be accelerated to reduce or eliminate slow crack growth. The test specimen is tested in hoop tension to fracture. The test specimen is retrieved for failure analysis and post-test dimensional measurement. A minimum of five valid tests is required for the purposes of estimating a mean. A greater number of tests may be necessary, if estimates regarding the form of the strength distribution are required. Fractography is suggested if the failure mode and fracture location are of interest.

CALCULATION, REPORTING, PRECISION AND BIAS

Calculations -- Using the measured force and or/pressure data along with the measured strain and/or deformation data as well as the test specimen dimensions, the resulting hoop stress-strain curve for each test specimen is determined. Calculation of the hoop stress is dependent on test method as shown in Fig. 9. Calculations for the elastomeric insert method may need to account for friction effects between the insert and the walls of the tubular test specimen. From the stress-stain curve, the following hoop tensile properties are determined: i) ultimate hoop tensile strength and corresponding strain, ii) fracture hoop tensile strength and corresponding strain, iii) proportional limit hoop tensile stress and corresponding strain, iv) elastic modulus in the circumferential direction, v) modulus of toughness.

$$\sigma_\theta = P\left[\frac{2r_i^2}{r_o^2 - r_i^2}\right] \text{ and } \varepsilon_\theta = \text{ measured directly}$$

[at outer radius for internal pressure]

where: P = pressure
 [internal hydraulic pressurization]

or

P = f (F$_{axial}$, A$_{insert}$, Elastic constants, stiffnesses)
 [elastomeric insert loading]

Figure 9 Calculation of hoop stress depending on pressurization method

Reporting -- The test methods provide detailed lists of reporting requirements for test identification, material and test specimen description, equipment and test parameters, and test results (statistical summary and individual test data).

Precision -- CMCs have probabilistic strength distributions, based on the inherent variability in the composite: fibers, matrix, porosity, fiber interface coatings, fiber architecture and alignment, anisotropy, and inherent surface and volume flaws. This variability occurs spatially within and between test specimens. Data variation also develops from experimental variability in test specimen dimensions, volume/size effects, extraneous bending stresses, temperature and humidity effects and the accuracy and precision of transducers and sensors.

Once the test methods are drafted, vetted and balloted, ASTM Committee C28 is planning interlaboratory testing programs per ASTM Practice E691[26] to determine the precision (repeatability and reproducibility) for a range of ceramic composites, considering different compositions, fiber architectures, and specimen geometries.

CURRENT STATUS AND FUTURE WORK

The draft standard test methods for hoop tensile behavior of CMC tubes are scheduled for first rounds of consensus ballots at subcommittee and main committee levels in late Spring 2013. If balloting is successful, publication of the full consensus standard test methods is expected for Fall 2013 or Spring 2014. Once the standards have been published, a round-robin interlaboratory testing program will be organized and executed, given available material, funding, and participating laboratories.

With sufficient interest and participation within the CMC community, new mechanical test standards for CMC tubes are planned for axial tensile strength, torsional shear strength, and flexural strength.

CONCLUSIONS

There is a real need for a comprehensive and detailed consensus test standard for hoop tensile testing of CMC tubes. This need is based on the certification and qualification requirements for CMC tubes in nuclear fission reactors. Test standards for tubes are needed because tests on flat composite panels are not representative of the architecture and geometry of composite tubes, with their 2-D and 3-D fiber architectures. The proposed ASTM standard test methods for hoop tensile testing of CMC tubes will be comprehensive and detailed, providing strong procedural documents using the conventional ASTM format. These new standard test methods will be applicable to 1-D, 2-D, and 3-D CMC tubes with diameters up to 150 mm and wall thicknesses up to 25 mm. The test methods will address the following experimental issues -- test specimen geometries and preparation, different loading methods, test equipment, interferences (material, specimen, parasitic stresses, test conditions, etc), testing modes and procedures, data collection, calculations, reporting requirements, and precision/bias.

ACKNOWLEDGEMENT
This work was conducted with U.S. Department of Energy funding under the technical direction of Dr. Yutai Katoh at Oak Ridge National Laboratory, Oak Ridge, TN.

REFERENCES
1. G. Griffith, "U.S. Department of Energy Accident Resistant SiC Clad Nuclear Fuel" INL/CON-11-23186, Idaho National Laboratory, Idaho Fall, Idaho (2011)

2. W. E. Windes, P. A. Lessing, Y. Katoh, L. L. Snead, E. Lara-Curzio, J. Klett, C. Henager, Jr., R. J. Shinavski, " Structural Ceramic Composites for Nuclear Applications," Idaho National Laboratory, Report INL/EXT-05-00652, Aug. (2005)

3. L.L. Snead, Y. Kato, W. Windes, R. J. Shinavski, T. Burchell, "Ceramic Composites For Near Term Reactor Application," Proceedings of the 4th International Topical Meeting on High Temperature Reactor Technology, HTR2008-58050, ASME International (2008)

4. M. G. Jenkins, E. Lara-Curzio, W. E. Windes, "(GENIV) Next Generation Nuclear Power And Requirements For Standards, Codes And Data Bases For Ceramic Matrix Composites," Ceramics in Nuclear and Alternative Energy Applications, Ceramic Engineering and Science Proceedings, Sharon Marra, ed., American Ceramics Society, Vol. 27, Issue 5, pp. 3-9 (2007)

5. Y. Katoh, L.L. Snead , T. Nozawa, N.B. Morley, W.E. Windes, "Advanced Radiation-Resistant Ceramic Composites," Advances in Science and Technology, Vol. 45, pp. 1915-1924 (2006)

6. ASTM D2290 - 08 "Standard Test Method for Apparent Hoop Tensile Strength of Plastic or Reinforced Plastic Pipe by Split Disk Method," ASTM International, West Conshohocken, PA (2013)

7. M. Rozental-Evesque, B. Rabaud, M. Sanchez, S. Louis and C-E. Bruzek, "The NOL Ring Test an Improved Tool for Characterising the Mechanical Degradation of Non-Failed Polyethylene Pipe House Connections" Plastic Pipes XIV, Budapest, Hungary (2008)

8. R. Carter, "Compressed Elastomer Method for Internal Pressure Testing", ARL-TR-3921, Aberdeen Proving Ground, MD (2006)

9. K. Mosley, "The Stressing for Test Purposes in Tubular Form Using Elastomeric Inserts-Experimental and Theoretical Development, " Proc. Instn Mech Engrs, Vol 196, pp. 123-139 (1982)

10. G.A. Graves and L. Chuck, "Hoop Tensile Strength and Fracture Behavior of Continuous Fiber Ceramic Composite (CFCC) Tubes from Ambient to Elevated Temperatures", J. Composites Technology and Research, Vol 19, No 3 (1997)

11. R.E. Ely, "Hoop Tension Strength of Composite Graphite-Aluminum Tubes", Army Missile Research Development and Engineering Lab Redstone Arsenal Al Physical Sciences Directorate, AD0489900, 24 Aug (1966)

12. T.R. Barnett, G.C. Ojard, and R.R. Cairo, "Relationships of Test Materials and Standards Development to Emerging Retrofit CFCC Markets, in Mechanical, Thermal and Environmental Testing and Performance of Ceramic Composites and Components, ASTM STP 1392, M.G. Jenkins, E. Lara-Curzio, S. T. Gonczy, eds. American Society for Testing and Materials, West Conshohocken, Pennsylvania (2000)

13. F. A. R. Al-Salehi, S. T. S. Al-Hassani, H. Haftchenari and M. J. Hinton, "Temperature and Rate Effects on GRP Tubes Under Tensile Hoop Loading" *Applied Composite Materials*, Vol. 8, No. 1 pp. 1-24 (2001)

14. H. Haftchenari, F. A. R. Al-Salehi, S. T. S. Al-Hassani and M. J. Hinton, "Effect of Temperature on the Tensile Strength and Failure Modes of Angle Ply Aramid Fibre (KRP) Tubes Under Hoop Loading," Applied Composite Materials, Vol. 9, pp. 99-115 (2002)

15. Emrah Salim Erdiller, "Experimental Investigation for Mechanical Properties of Filament Wound Composite Tubes," Thesis, Middle East Technical University, Ankara, Turkey May 2004

16. M. J. Verrilli, J. A. DiCarlo, H.M., Yun, T. R. Barnett, "Hoop Tensile Properties of Ceramic Matrix Composites, *J. of Testing and Evaluation*, Vol. 33, No. 5 (2005)

17. Pinar Karpuz, "Mechanical Characterization Of Filament Wound Composite Tubes By Internal Pressure Testing" Thesis, Middle East Technical University, Ankara, Turkey, May 2005

18. J.A. Salem and J.Z. Gyekenyesi, "Burst Pressure Testing of Ceramic Rings," NASA/TM-2007-214695, April 2007.

19. J. Cain, S. Case, and J. Lesko, J., "Testing of Hygrothermally Aged E-Glass/Epoxy Cylindrical Laminates Using a Novel Fixture for Simulating Internal Pressure." *J. Compos. Constr.*, Vol. 13, No. 4, pp. 325–331 (2009).

20. G. D. Roberts, J. A. Salem, J. L. Bail, L.W. Kohlman, W. K. Binienda, and R. E. Martin, "Approaches For Tensile Testing of Braided Composites," 5th International Conference on Composites Testing and Model Identification (Comp Test 2011), 14-16 February 2011, Lausanne, Switzerland

21. R. Rafiee, "Apparent hoop tensile strength prediction of glass fiber-reinforced polyester pipes," *J. Composite Materials*, first published on May 22, 2012 as doi:10.1177/0021998312447209.

22. M. Robert and A. Fam, "Long-Term Performance of GFRP Tubes Filled with Concrete and Subjected to Salt Solution," *J. Compos. Constr.*, Vol. 16, No. 2 , pp. 217-224 (2012)

23. ASTM D 1599-99 "Standard Test Method for Resistance to Short-Time Hydraulic Pressure of Plastic Pipe, Tubing, and Fittings," ASTM International, West Conshohoken, PA (2013)

24. K. Gramoll, "Multimedia in Engineering," Thesis, University of Oklahoma, Norman, Oklahoma (2011)

25. S.T. Gonczy and M.G. Jenkins, "A New ASTM C28 Test for the Uniaxial Tensile Properties of Carbon-Carbon and SiC-SiC Composite Tubes", *Ceramic Engineering and Science Proceedings* (2012)

26. ASTM E691-99 "Standard Practice for Conducting an Interlaboratory Study to Determine the Precision of a Test Method," ASTM International, West Conshohoken, PA (2013)

TEST METHODS FOR FLEXURAL STRENGTH OF CERAMIC COMPOSITE TUBES FOR SMALL MODULAR REACTOR APPLICATIONS

Michael G. Jenkins, Bothell Engineering & Science Technologies, Fresno, CA, USA

jenkinsmg@bothellest.com

Thomas L. Nguyen, Levitas Consultants, Merced CA, USA

tln@levitas.net

ABSTRACT

US DOE is planning to use advanced materials for the core and the reactor unit components in various advanced Small Modular Reactor (SMR) concepts. Ceramic matrix composites (CMC), in particular silicon carbide (SiC) fiber SiC-matrix (SiC/SiC) composites, could revolutionarily expand the design window for various components in terms of operating temperature, applicable stress, and service life, as compared to heat-resistant metallic alloys, while significantly improving safety margins and accident tolerance. Examples of CMC tubular components include control rod sleeves, control rod joints, and fuel rods. Anticipated failure modes for these components include axial and hoop tension, axial flexure, axial and diametral compression, and axial shear. Because there are no commonly-accepted design methodologies for advanced composite tubular components, there are almost no mechanical test standards for any properties of CMC tubular components. Therefore, some current and proposed test methods for measuring axial-flexure strength of composite tubes are presented, discussed, and compared for application to CMCs. A proposed test method is presented in terms of the following experimental issues -- test specimen geometries/preparation, test fixtures, test equipment, interferences, testing modes/procedures, data collection, calculations, reporting requirements, precision/bias.

KEYWORDS – axial flexure strength, tubes, CMC, nuclear fission.

INTRODUCTION

The US Department of Energy (US DOE) is accelerating the timelines for the commercialization and deployment of small modular reactors (SMRs) as part of its strategy for development of clean, affordable nuclear power options. As a part of this strategy, a high priority has been to help technologies advance the licensing and commercialization of domestic SMR designs that are relatively mature and can be deployed in the next decade.

SMRs offer the advantage of lower initial capital investment, scalability, and siting flexibility at locations unable to accommodate larger reactors. In addition, SMRs have enhanced safety and security through the following attributes[1]:

Modularity: SMRs have the advantage that major components of the nuclear steam supply system can be fabricated in a factory and shipped to the point of use. Therefore, SMRs require limited on-site preparation and can reduce lengthy construction times that are typical of the larger units. SMRs are simpler in design, contain enhanced safety features, benefit from the economics and quality of factory production, and provide for more flexibility.

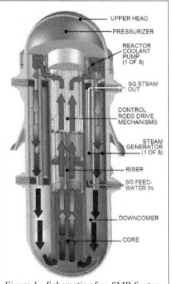

Figure 1 - Schematic of an SMR System
(from Ref. 1)

Lower Capital Costs: SMRs can reduce the investment of a nuclear plant owner because of the lower plant capital cost. Modular components and factory fabrication can reduce construction costs and construction times.

Site Flexibility: SMRs provide power for applications where large plants are not needed or sites that lack infrastructure such as small electrical markets, isolated regions, small electrical grids, sites with limited water and acreage, or unique industrial applications. SMRs are attractive for either replacement or repowering of outdated fossil-fuel plants. SMRs may even complement existing industrial processes or power plants with a non-greenhouse gas-energy source.

Gain Efficiency: SMRs, when coupled with other energy sources, including renewable and fossil energy, can produce higher efficiencies and multiple energy end-products while increasing grid stability and security. Some proposed SMR designs produce high-temperature process heat for industrial applications, including generation of electricity.

SMRs designed from advanced and innovative concepts, using non-LWR coolants such as liquid metal, helium or liquid salt, may offer added functionality and affordability. To achieve the goal of the SMR program, activities are focusing on four key areas[1]:

- Developing assessment methods for evaluating advanced SMR technologies and characteristics;
- Developing and testing of materials, fuels and fabrication techniques;
- Resolving key regulatory issues identified by US Nuclear Regulatory Commission (NRC) and industry; and
- Developing advanced instrumentation and controls and human-machine interfaces.

In this regard, non-metallic fiber-reinforced composites [e.g., carbon-carbon (C-C) and silicon carbide-silicon carbide (SiC-SiC)] are being developed as primary structural materials in the hot-core sections of the SMRs. These CMCs have high strength, high toughness, low coefficient of thermal expansion, excellent resistance to thermal creep, and superior resistance to oxidation and corrosion at high-temperatures.[2, 3] A key application for CMCs is for the control rods in the core. The design concept for the control rod uses a flexible string of rigid CMC tubes containing boronated graphite compacts. The sections of CMC tube control-rods are 500 mm long and ~60 mm in diameter.

The CMC tubes must be stable and damage resistant under long-term high flux neutron irradiation in the fission core. The CMCs consist of high-strength graphite and silicon carbide fibers in a high temperature ceramic matrix (graphite or silicon carbide), providing high strength and high fracture resistance at elevated temperatures. The ceramic reinforcement tows have high filament counts (500-2000) and are woven with large units cells, several millimeters in size. In a tube configuration, the composites can have a 1-D filament wound, 2-D laminate, or 3-D (weave or braid) construction depending on what tensile, shear, and hoop stresses are considered. The fiber architecture in the tubes can be geometrically tailored for highly anisotropic or uniform isotropic mechanical and thermal properties.[3,4] SiC-SiC composites are considered as primary materials, because they have greater resistance to neutron radiation than C-C composites. [5,6]

Tubular geometries for nuclear applications present challenges for both "makers" and "lookers" of SiC/SiC CMCs. For "makers": how to make seamless tubes with multiple direction architectures; how to ensure integrity in the radial direction; how to create uniform wall thickness and uniform/nonporous matrices. For "lookers": how to build on decades of experience with consensus standards and data bases for "flat" material forms; how to interpret information of tests of test specimen in component form; how to adapt expertise at room temperature in ambient environments to conditions at high temperature in specific extreme-use environments.

The NRC has stringent requirements for the regulatory acceptance, qualification, certification, and licensing of materials in nuclear reactors.[2,5] These requirements cover the design, qualification, production, installation, and operation of these new materials in a reactor.

Fortunately the nuclear applications of SiC/SiC CMCs builds on experience allowing nuclear applications to advance an existing mature specialized technology. For example, SiC/SiC CMC materials and structure technology were funded by the aerospace and defense industries/agencies. In addition, current evaluations and applications of SiC/SiC CMCs in fusion reactors (first wall) and tristructural-isotropic (TRISO) fuel forms that have established properties under extended neutron irradiation and at high temperatures as well as very hot steam environment. Growing, credible data bases for SiC/SiC CMCs now exist because of the evolution of consensus test methods and design codes. Finally, maturation of volume-scale manufacturing capability for all types of CMCs including SiC/SiC CMC adds to availability and understanding of these material.

Such professional organizations as ASME and ASTM are taking the lead in developing the codes, specifications, and test standards for CMCs in nuclear applications. ASTM Committee C28 on Advanced Ceramics has a particular focus on mechanical test standards for CMCs. Specifically, ASTM Subcommittee C28.07 has published eleven standards for CMCs (e.g., tensile, flexure, shear, compression, creep, fatigue, etc.

Mechanical testing of composite tube geometries is distinctly different from testing flat plates because of the differences in fiber architecture (weaves, braids, filament wound), stress conditions (hoop, torsion, and flexure stresses), gripping, gage section definition, and scaling issues.[5] Because there are no commonly-accepted design methodologies for advanced composite tubular components, there are almost no mechanical test standards for any properties of tubular ceramic composite components. Therefore, in this paper, some current and proposed test methods for measuring flexural strength of composite tubes are presented, discussed, and compared for application to CMCs. A proposed draft test method is presented in terms of the following experimental issues -- test specimen geometries/preparation, test fixtures, test equipment, interferences, testing modes/procedures, data collection, calculations, reporting requirements, precision/bias.

FLEXURAL STRENGTH TESTS OF TUBES

A review of the literature for experimental and analytical methods applied to assessing behavior of tubes subjected to flexural stress resulted in the following categories.

1) Transverse loading of tubes simply supported as beams in bending (3 and 4-point loading)
2) Cantilever-loaded tubes (sometimes over a bend die)
3) Tubes subjected to moments

Aspects of each category are discussed and illustrated in the following sub sections.

Figure 2 Illustration of actual 3-point and 4-point bend tests of composite tubes. Note crushing at central load point for the 3-point loading configuration (from Refs 7 and 8).

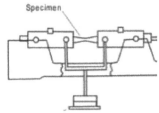

Figure 3 Illustration of schematic illustration of 4-point bend tests of using collet gripped circular test specimen as applied to rotating bending fatigue testing (from Ref 9).

Transverse loading of tubes simply supported as beams in bending (3 and 4-point loading)-Long sections of tubes are loaded transversely as beams in either 3 or 4 point bending as illustrated in Fig. 2. While conceptually simple and straightforward to implement, a major drawback is crushing or failure initiated at load points, often due to the point contact of loading rollers and the curvature of the tube. A possible to eliminate wall crushing to use a method successfully applied to rotating bending fatigue testing of cylindrical test specimens held in the bend fixtures by collet grips. If applied to flexure testing CMC tubes, this type of arrangement could use the same plug and collet system developed for tensile testing CMC and crushing of the tube at the transverse load points would be eliminated. Table 1 shows some pros and cons for this type of test classification.

Cantilever-loaded tubes-This is a simple method based on tube bending as shown in Fig. 4. For flexure testing of composite tubes, the mandrel can be replaced by end plugs and collets that have been used in tensile testing composite tubes. The bend die can be chosen to fit the tube characteristics. Common problems are often related to failures initiated at the interface between the plug and the tube or at the interface of the tube and the bend die. The simplicity of the loading scenario and availability of testing fixtures are advantages. Table 2 shows some pros and cons for this type of test classification.

Table 1 Some Pros and Cons for Transverse Loading of Tubes Simply Supported as Beams in Bending

Pros	Cons
- "Simple" fixtures for 3-4 point loading	- Potential wall crushing at load points
- Similar collet system to tensile tests for constant moment loading	- Interpretation of results
- Uses existing test machines and instrumentation	- Mixed failure modes (wall buckling in compression and fibre failure in tension)
- Simple equations	

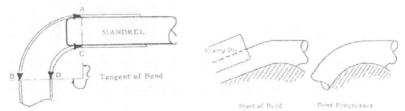

Figure 4 Illustration of cantilever flexure test of composite tube (i.e. tube bending) (from Ref 10).

Tubes subjected to moments-This is a conceptually simple but experimentally challenging method for applying bending moments to tubes. Pure bending moments can be used to easily calculate the flexure stress in the tube. However, achieving pure bending can be difficult. One concept is to attach moment arms to the plugs and collets used in tensile testing tubes and apply axial loads to the end of the moment arms as shown in Fig. 5. The magnitude of the resulting moment is related to the length of moment arms and level of the axial force. A complication of this method is the "purity" of the applied bending moment. Table 3 shows some pros and cons for this type of test classification.

Table 2 Some Pros and Cons for Cantilever-Loaded Tubes (sometimes over a bend die)

Pros	Cons
- Simple fixtures	- May be complex stress states
- Uses short or long sections of tube	- Friction effects for rough surfaces
- Uses existing test machines	- Potential for non representative failures
- Practical experience base	- May be limited to material selection

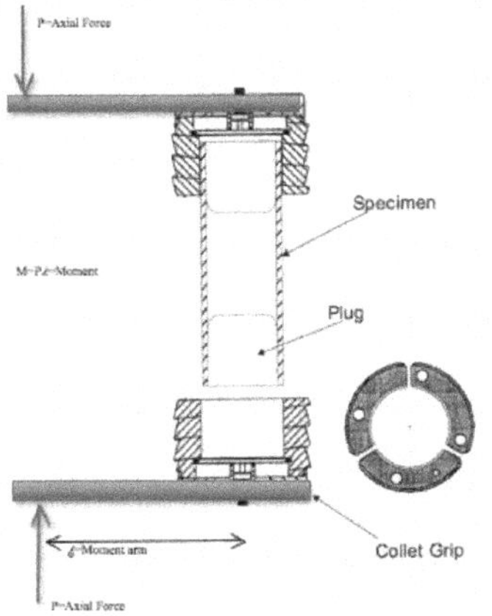

Figure 5 Illustration of bending moment applied to composite tube by axial loading

Table 3 Some Pros and Cons for Tubes Subjected to Moments

Pros	Cons
- Uses existing collets and plugs	- Combined stresses (axial+bending)
- Applies moment axially	- Complex end collets/plug/arm
- Uses existing test machines	- Mixed failure modes (wall buckling in compression and fibre failure in tension)

The complexity of the nonhomogeneous, anisotropic, nonlinear elastic behaviour of fibre-reinforced composites has lead to differences in results from the different methods discussed here. Therefore, the most appropriate test methods for ceramic composites at room temperature are those that result in bending stresses that arise from application of moments. The evolving standardized test method for flexural strength of ceramic composites described in the following sections reflects this conclusion.

CMC TUBE FLEXURAL TEST -- SCOPE AND APPLICATION

A working group of ASTM Subcommittee C28.07 on CMC tube testing is developing a draft test method: "Flexural Behavior of Continuous Fiber-Reinforced Advanced Ceramic Composite Tubular Test Specimens at Ambient Temperature." In the test method a composite tube/cylinder with a defined gage section and a known wall thickness is fitted/bonded into a loading fixture. The test specimen/fixture assembly is mounted in the testing machine and monotonically loaded in at ambient temperature while recording the applied force. Ultimate flexural strength and fracture flexural strength along with corresponding strains are determined from the maximum applied force and the fracture force, respectively, the resulting moments and the test specimen dimensions. Proportional limit stress and the axial modulus of elasticity in flexure are determined from the stress-strain data.

The test method addresses test equipment, gripping methods, testing modes, interferences, tubular test specimen geometries, test specimen preparation, test procedures, data collection, calculation, reporting requirements, and precision/bias. It is applicable to a wide range of CMC tubes with 1-D filament, 2-D laminate, and 3-D weave and braid architectures. In addition, where appropriate, the test method references test procedures and fixturing from research work done on CMC and PMC tubes.

EXPERIMENTAL FACTORS

CMCs generally exhibit "graceful" failure from a cumulative damage process, unlike monolithic advanced ceramics that fracture catastrophically from a single dominant flaw. The flexure testing of CMC (both flats and tubes) has a range of different material and experimental factors that interact and must be controlled and managed:

• Material Variability, including Anisotropy, Porosity, and Surface Condition • Test Specimen Size, Fiber Architecture, and Gage Section Geometry Effects • Gripping and Bonding Failures	• Out-Of-Gage Failures and Extraneous Stresses • Slow Crack Growth, Strain Rate Effects and Test Environment • Accurate Strain/Elongation Measurement

Fig. 6 – Schematic of Straight-Sided Tubular Test Specimen (from Ref 11)

Fig. 7 – Schematic of Contoured Gage Section Tubular Test Specimen (from Ref 11)

All of these factors must be managed and understood to produce consistent, representative flexure failure in the gage section of the test specimens. Tubular test specimens with cylindrical geometries provide particular challenges in the areas of gage section geometry, gripping and bonding failures, extraneous stresses, and out-of-gage failures.

TEST SPECIMEN GEOMETRIES

Test Specimen Size --CMC tubes are fabricated in a wide range of geometries and sizes, across a spectrum of fiber-matrix-architecture combinations. Defining a single test specimen geometry that is universally applicable may not be practical. The selection and definition of a test specimen geometry depends on the purpose of the flexural testing effort. With that consideration, the test method is generally applicable to tubes with outer diameters (D_o) of 10 to 150 mm and wall thicknesses (t) of 1 to 25 mm, where the ratio of the outer diameter to wall thickness is commonly D_o/t = 5 to 30. In many cases, the wall thickness is defined by the fiber-reinforcement architecture, particularly for woven and braided configurations.

Gage Section Geometry – Tubular test specimens are classified into two groups – straight-sided and contoured, as shown in Figures 6 and 7. Contoured gage-section test specimens are distinctive in having gage sections with thinner wall thicknesses than in the grip sections.

Although straight-sided test specimens are easier to fabricate and are commonly used, tubular test specimens with contoured gage sections are preferred to promote tensile failure in defined gage section. The contoured gage sections are formed by one of two methods – 1) integral thick-wall grip sections in the composites or 2) adhesively-bonded collars/sleeves (OD and/or ID) in the grip sections as shown in Figures 8 and 9, respectively.

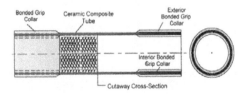

Fig. 8 – Adhesively Bonded Collars for Gripping (Not to Scale) (from Ref 11)

Experience has shown that successful tests can be maximized by using consistent ranges of relative gage section dimensions, as follows:

$$2 < L_O / D_o < 3 \qquad and \qquad 15 < L_O / t < 30 \qquad (1)$$

where L_O is the defined gage length, D_o is the outer diameter in the gage section, and t is the wall thickness in the gage section of the tube. Deviations from the recommended geometries may be necessary depending upon the particular composite tube geometry being evaluated.

As a starting point, the wall thickness of the non gage section should be at least twice as thick as the wall thickness of the gage section. As a general rule, non gage section lengths are >1.5 times the outer diameter of the test specimen. If the test specimen fails in the non gage section, longer non gage sections may be needed. A key factor in contoured gage-section test specimens is minimizing stress concentrations at the geometric transitions into the gage sections, using tapered or rounded transitions.

TEST EQUIPMENT AND PROCEDURES
    Loading Mechanism -- Various types of loading devices may be used to apply the measured force from the testing machine to the tubular test specimens.

    Passive type of grip fixtures transmit the force applied by the test machine to the tubular test specimen through a direct adhesive bond into the grips or by mechanical action between geometric features on the test specimen and the grip fixture. An example of a bonded grip system is shown in Figure 9.

    Insufficient bonding surface in the adhesive-bonded non gage section may produce bond failure before test specimen failure. As a rule of thumb the bond shear forces that develop from the maximum force should produce shear stresses <50% of the nominal shear strength of the adhesive. In bonded systems, one of the concerns is the removal of the test specimen from the grips after testing. The adhesive must be removed by either chemical or thermal action, depending on the nature of the adhesive.

    Load Frame, Force and Strain Measurement, Data Acquisition -- The test method uses a standard load frame with a hydraulic or screw drive loading mechanism and standard force transducers. Load is applied axially as shown in Figs 4. Extension/strain is measured first by extensometers along the gage length, with an option to use bonded resistance strain gage rosettes to measure both axial and transverse strains. If required, an environmental test chamber may be used to control humidity and ambient temperature. Data collection should be done with a minimum of 50-Hz response and an accuracy of ±0.1 % for all data.

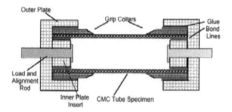

Fig. 9 – Schematic of an adhesive-bonded non gage section for a straight-sided tube (from Ref 10)

Test Procedures – Generally, a displacement-control or strain-control test mode is used to avoid "run-away" tests that sometimes occur in force-control. A test mode rate is chosen so as to produce test specimen failure in 5-50 s. Failure within 1 minute or less should be sufficient to minimize slow-crack growth (SCG) effects in the tensile test. If slow crack growth is observed (e.g. under slow test mode rates), subsequent tests can be accelerated to reduce or eliminate slow crack growth. Extensometers and/or strain gages are used to determine gage length extension and/or local strain. The test specimen is tested in flexure to fracture. The test specimen is retained for failure analysis and post-test dimensional measurement.

A minimum of five valid tests is required for the purposes of estimating a mean. A greater number of tests may be necessary, if estimates regarding the form of the strength distribution are required.

CALCULATION, REPORTING, PRECISION AND BIAS

Calculations -- Using the measured force-displacement data and resulting calculated stress-strain data along with test specimen dimensions, the following flexural properties are determined: i) ultimate flexural strength and corresponding strain, ii) fracture flexural strength and corresponding strain, iii) proportional limit flexural stress, iv) elastic modulus in flexure.

Reporting -- The test method provides a detailed list of reporting requirements for test identification, material and test specimen description, equipment and test parameters, and test results (statistical summary and individual test data.)

Precision -- CMCs have probabilistic strength distributions, based on the inherent variability in the composite: fibers, matrix, porosity, fiber interface coatings, fiber architecture and alignment, anisotropy, and inherent surface and volume flaws. This variability occurs spatially within and between test specimens. Data variation also develops from experimental variability in test specimen dimensions, volume/size effects, extraneous bending stresses, temperature and humidity effects and the accuracy and precision of transducers and sensors.

ASTM Committee C28 is planning interlaboratory testing programs per ASTM Practice E691[12] to determine the precision (repeatability and reproducibility) for a range of CMCs, considering different compositions, fiber architectures, and specimen geometries.

CURRENT STATUS AND FUTURE WORK

The proposed draft standard test method for flexure behavior of CMC tubes is scheduled for a first round of consensus ballots at subcommittee and main committee levels in the late Spring 2013. If balloting is successful, publication of the full consensus standard test method is expected for Fall 2013 or Spring 2014. Once the standard has been published, a round-robin interlaboratory testing program will be organized and executed, given available material, funding, and participating laboratories.

CONCLUSIONS

There is a real need for a comprehensive and detailed consensus test standard for flexure testing of CMC tubes. This need is based on the certification and qualification requirements for CMC tubes in nuclear fission reactors. A test standard for tubes is needed because tests on flat composite panels are not representative of the architecture and geometry of composite tubes, with their 2-D and 3-D fiber architectures. A proposed ASTM draft standard test method for flexure testing of CMC tubes is comprehensive and detailed, providing a strong procedural document using the conventional ASTM format. This proposed standard test method is applicable to 1-D, 2-D, and 3-D CMC tubes with diameters up to 150 mm and wall thicknesses up to 25 mm. The test method addresses the following experimental issues -- test specimen geometries and preparation, different gripping methods, test equipment, interferences (material, specimen, test conditions, etc.), testing modes and procedures, data collection, calculation, reporting requirements, and precision/bias.

ACKNOWLEDGEMENT
This work was done with U.S. Department of Energy funding under the technical direction of Dr. Yutai Katoh at Oak Ridge National Laboratory, Oak Ridge, TN

REFERENCES
1. Small Modular Nuclear Reactors, US Department of Energy, Washington, DC
http://energy.gov/ne/nuclear-reactor-technologies/small-modular-nuclear-reactors, February (2013)

2. NGNP High Temperature Materials White Paper, Sec 3.5 Composite Materials, Idaho National Laboratory, INL/EXT-09-17187, June (2010)

3. W. E. Windes, P. A. Lessing, Y. Katoh, L. L. Snead, E. Lara-Curzio, J. Klett, C. Henager, Jr., R. J. Shinavski, " Structural Ceramic Composites for Nuclear Applications," Idaho National Laboratory, Report INL/EXT-05-00652, Aug. (2005)

4. L.L. Snead, Y. Katoh, W. Windes, R. J. Shinavski, T. Burchell, "Ceramic Composites For Near Term Reactor Application," Proceedings of the 4th International Topical Meeting on High Temperature Reactor Technology, HTR2008-58050, ASME International (2008)

5. M. G. Jenkins, E. Lara-Curzio, W. E. Windes, "(GENIV) Next Generation Nuclear Power And Requirements For Standards, Codes And Data Bases For Ceramic Matrix Composites," Ceramics in Nuclear and Alternative Energy Applications, Ceramic Engineering and Science Proceedings, Sharon Marra, ed., American Ceramics Society, Vol. 27, Issue 5, pp. 3-9 (2007)

6. Y. Katoh, L.L. Snead, T. Nozawa, N.B. Morley, W.E. Windes, "Advanced Radiation-Resistant Ceramic Composites," Advances in Science and Technology, Vol. 45, pp. 1915-1924 (2006)

7. Bryce Ingersoll "Development of Flexure Testing Fixtures and Methods for Thin-Walled Composite Tubes" Thesis, University of Utah, Salt Lake City, UT (2011)

8. Puneet Saggar "Experimental Study Of Laminated Composite Tubes Under Bending" Thesis, University of Texas at Arlington, Arlington, TX (2007)

9. ASM International, Chapter 14 Fatigue in Elements of Metallurgy and Engineering Alloys, ASM International, Materials Park, OH (2013)

10. "Basic Tube Bending Guide," Hines Bending Systems, Fort Meyers, FL (2012).

11. S.T. Gonczy and M.G. Jenkins, "A New ASTM C28 Test for the Uniaxial Tensile Properties of Carbon-Carbon and SiC-SiC Composite Tubes", Ceramic Engineering and Science Proceedings (2012)

12. ASTM E691-99 "Standard Practice for Conducting an Interlaboratory Study to Determine the Precision of a Test Method," ASTM International, West Conshohocken, PA (2013).

EFFECTS OF SIZE AND GEOMETRY ON THE EQUIBIAXIAL FLEXURAL TEST OF FINE GRAINED NUCLEAR GRAPHITE

Chunghao Shih[1], Yutai Katoh[1], Takagi Takashi[2]
[1]: Oak Ridge National Laboratory, Oak Ridge, TN, United States
[2]: Ibiden, Gifu, Japan

ABSTRACT

Reduction of specimen size is an essential requirement for neutron irradiation effect studies of materials since space in nuclear reactors for those studies are often very limited and costly. Reducing the specimen size can dramatically reduce the cost of these studies, the irradiation exposure to personnel involved in post irradiation examination and the volume of radioactive waste.

In this study, 4 specimen sizes were systematically varied to examine the size effect on the equibiaxial flexure test of fine grain nuclear graphite in non-irradiated conditions at room temperature. A statistically significant sample population of 30 was used for each specimen geometry or dimension.

The results showed that the round shaped samples followed weibull distribution with Weibull modulus of 25 to 35. The average strength followed the trend of the weakest link theory. The 2 sizes of square shaped samples had Weibull modulus of 29 and 40. Data from round and square samples agree with others reasonably and collectively follow the Weibull scaling theory. It is concluded that the equibiaxial flexural test using adequate population of small specimens provides useful and reliable information on the statistical strength properties for fine-grained nuclear graphite materials.

INTRODUCTION

Graphite is used as structure materials and moderators in nuclear reactors. The mechanical and physical properties of nuclear grade graphite following irradiation damage have been extensively studies[1,2,3]. Neutron irradiation initially produces significant volume shrinkage together with increase in strength and modulus. With increasing fluence, the graphite will commence volume expansion and eventually the volume will return to its pre-irradiation value, which is considered to be the end of service lifetime for the material. As graphite's volume expands, the strength and modulus will also decrease accordingly.

Four point bend flexural test is the most common method to determine the fracture strength of graphite materials and has the advantage of simple configuration and high temperature compatible[4]. However, this type of test generally requires specimens with substantial volume (typically of at least several cubic centimeters per specimen). Moreover, due to the statistical nature of graphite strength, a sample population of over 30 specimens and Weibull statistical analysis are required, resulting in large material demand in order to produce statistically significant strength data. This disadvantage makes developing a flexural strength test method with miniaturized specimens attractive, especially for irradiation studies, where specimen volume and radiation exposure are a major concern.

The equibiaxial flexural test is a promising technique with the following advantages: 1) small specimen can be utilized, 2) test configuration is simple, and 3) the effects of edge defects are eliminated. The aim of this work is to investigate the specimen size and geometry effects on the equibiaxial flexural test of fine grained nuclear graphite. It is hoped to understand those effects in order to confidently utilize the miniaturized specimens with this test method for the irradiation studies.

EXPERIMENTAL

Fine grained nuclear grade graphite (grade ETU-10) from Ibiden Japan was used in this study. It has a density of 1.76 g/cm³ and a maximum manufacture claimed grain size of 40 μm. The graphite was machined into 4 different sizes of round disks and 2 different sizes of square disks to study the specimen size and shape effect on the equibiaxial flexural test. Table I shows the detailed dimension of those samples. The specimens were machined to a surface roughness of less than 3 μm Ra. No further polishing of the specimen surfaces was done before fracture tests. A thorough weight and dimensional inspection was done for each specimen to eliminate any specimens with material or machining imperfections. Equibiaxial flexural strength of those specimens was determined following the guidelines of ASTM C1499-09. The dimensions of the fixtures used for different sizes of specimens are shown in Table I. Both the loading ring and the supporting ring were made of stainless steel. The actual ring diameters were measured with a Keyence VHX-1000E microscope (Keyence, Osaka, Japan). A Teflon alignment block was used to align the sample and the loading ring. A universal test frame (Insight 10, MTS, Minnesota, USA) was used for the flexural test with a cross head speed of 5 mm/min.

Four point bend test was also conducted per ASTM C651-11 to compare the determined strength values with the equibiaxial strength. The specimens were rectangular bars of 45mm × 4mm × 3mm. Two types of graphite orientations were tested in the four point bend test: one had the graphite extrusion axial parallel to the specimen length direction (axial) the other had the graphite extrusion axial parallel to the specimen width direction (transverse). The supporting span and loading span were 38.2mm and 12.8 mm, respectively. The crosshead speed was 1 mm/min.

30 replications were conducted for each specimen geometry or dimension in order to obtain a meaningful statistical analysis.

Table I. Specimen and fixture dimensions (all dimensions in mm)

Specimen dimension			Fixture dimension		Remark
Shape	D[1] or L[2]	Thickness	D_S[3]	D_L[4]	
○	6	0.5	5	2	BiAx D6×0.5
○	12	1	10	4	BiAx D12×1
○	25	2	20	8	BiAx D25×2
○	50	4	40	16	BiAx D50×4
□	6	0.5	5	2	BiAx 6×6×0.5
□	25	2	20	8	BiAx 25×25×2
Rec[5] (axial)	45×4×3		38.2	12.8	Ax. 4Pt 45×4×3
Rec[5] (transverse)	45×4×3		38.2	12.8	Tr. 4Pt 45×4×3

[1]: diameter
[2]: length
[3]: supporting ring diameter/supporting span
[4]: loading ring diameter/loading span
[5]: rectangular

RESULTS AND DISCUSSION

Average Flexural Strength

The average flexural strength of samples with different sizes, geometries and test methods is listed in Table II and graphically summarized in Figure 1. Representative specimen photographs after the equibiaxial flexural test are shown in Figure 2. The photographs show views from the compression side for clear vision of the loading ring trace. Dash lines are added when the loading ring trace is un-resolvable.

It is worth noting that in Kondo's report[4], severe stress concentration near the loading ring contact area was observed, both from fracture pattern observation and finite element analysis, because of the thinner thickness and stronger fracture strength of the pyro-carbon specimens in that study, which both caused larger deflection near the loading ring contact area. In this study, specimens have greater thickness and weaker fracture strength, which can mitigate the stress concentration near the loading ring contact area during the fracture test. Indeed, the fracture patterns in Figure 2 indicate that for all specimen sizes and geometries, crack initiated within the loading ring area, but not on or along the loading ring, which verifies the validity of the test.

For equibiaxial flexural test with round samples, minor size effect can be seen with larger samples having smaller average apparent flexural strength. The one exception is the BiAx D12×1 samples, which showed lower than expected strength values and larger standard deviation than other samples. The reason is unclear but is speculated to be either machining related defects or defects associated with the ring on ring fixture. For equibiaxial flexural test with square samples, no obvious size effect was observed considering the variation of the data. It is worth noting that more specimen alignment related errors are expected for square samples than round samples since the square samples are sensitive to machining imperfection of the specimens and the machining errors of the Teflon alignment block. These types of errors probably contributed to the absence of any observed size effect with the square samples.

Table II. Strength and Weibull modulus of different graphite samples

Specimen Type	Average strength (MPa)	Std.Dev.[1]	Weibull modulus	95% LB[2]	95% UB[3]
BiAx D50×4	62.6	2.5	30.7	23.0	41.3
BiAx D25×2	70.0	3.0	37.7	28.3	50.4
BiAx D12×1	62.1	5.4	12.8	9.6	17.1
BiAx D6×0.5	74.3	2.4	30.7	23.6	42.1
BiAx 25×25×2	71.6	2.1	37.8	28.7	50.7
BiAx 6×6×0.5	68.8	2.7	28.7	21.7	37.9
Ax. 4Pt 45×4×3	60.3	2.0	35.5	26.7	47.9
Tr. 4Pt 45×4×3	53.5	1.9	31.4	23.8	42.7

[1]: Standard deviation
[2]: 95% confidence lower bound
[3]: 95% confidence upper bound

No significant difference between the flexural strength values of the round and square sample geometry was observed. All the samples showed equibiaxial flexural strength of 62-75 MPa.

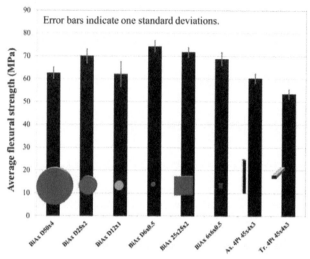

Figure 1. Average flexural strength of samples with different sizes, geometries and test methods

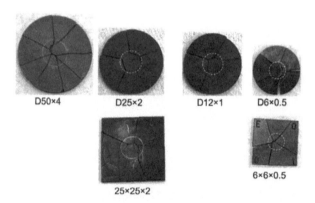

Figure 2. Representative photos of specimens after equibiaxial flexural tests

Four point bend test results indicate that sample orientation affects flexural strength. The axially oriented samples showed a higher flexural strength of 60.3 MPa while the transversely oriented samples showed a lower flexural strength of 53.5 MPa. The fine grained graphite (grade ETU-10) studied here showed higher flexural strength than other nuclear grade graphite, e.g. NBG-18[5] and PPEA[6] graphite, which both showed a flexural strength of ~30 MPa.

Weibull Analysis

The Weibull plot of flexural strength of all the samples is shown in Figure 3. The weibull modulus of flexural strength of all the samples is shown in Table II and graphically summarized in Figure 4. All the data sets showed good agreement with weibull statistical model, i.e. a good linear fit of each data set is seen in Figure 3. All the samples showed Weibull modulus of 30-38 except the BiAx D12×1 samples, which showed a lowered Weibull modulus of 12.8. This finding further confirms the previous speculation that the BiAx D12×1 data sets suffered from systematic defects from either machining or the ring on ring fixture. This data set is thus eliminated from the Weibull scaling analysis.

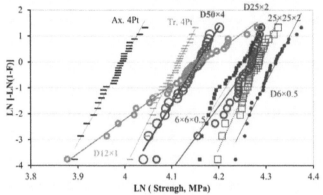

Figure 3. Flexural strength Weibull plot of samples with different sizes, geometry and test method

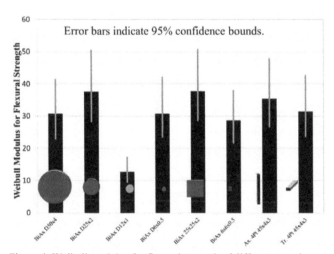

Figure 4. Weibull modulus for flexural strength of different samples

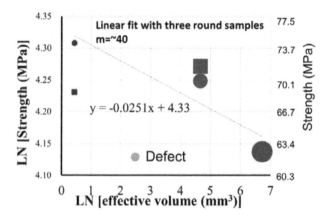

Figure 5. Flexural strength versus effect volume for sample with different sizes and shapes

Weibull's weakest link theory was used to investigate the size effect of the equibiaxial flexural test. The volume of the samples that lies within the loading ring diameter during the test was used as the effective volume. The theory leads to a strength dependency on effective volume:

$$\frac{\sigma_1}{\sigma_2} = (\frac{V_{E2}}{V_{E1}})^{\frac{1}{m}}$$ (1)

where σ_1 and σ_2 are the mean (or characteristic) strength of specimens of type 1 and 2, V_{E1} and V_{E1} are effective volume and m is the Weibull modulus. Equation 1 indicates that on the log strength versus log effective volume plot, the data points should fit into a straight line and the slope of the line equals -1/m. Such a plot for the equibiaxial flexural strength of the round and square shaped samples is shown in Figure 5. As discussed previously, the square samples don't follow the Weibull weakest link theory probably because of sample alignment issues. The round samples follow the theory except BiAx D12×1 data sets, which suffered from systematic defect. The linear regression of the three round sample data sets indicates a Weibull modulus of 39.8.

If surface defects are assumed to be the critical defect, the V (volume) in equation 1 should be replace by S (surface area). Plot similar to Figure 5 can be generated using effective surface area instead of effective volume and is presented in Figure 6. The obtained weibull modulus for the three round samples appears to be ~27. With the present results, it is difficult to determine whether the critical flaws reside on the specimen surface or inside the specimen volume. It is speculated that the critical flaw location is a combination of both cases among all the specimens tested.

Both volume and surface flaw assumptions result in Weibull modulus values that are reasonably close to the Weibull modulus obtained from the 30 replicates of the same size of samples as shown in Table II (~30). However, the critical surface flaw assumption might lead to a better fit since the obtained Weibull modulus of ~27 is closer to that obtained from individual data sets. This agreement suggests that Weibull statistical model is applicable for equibiaxial flexural strength across different specimen sizes for the round shaped samples. Small size samples can thus be used to evaluate flexural strength of fine grained nuclear graphite. The reduction in sample size can dramatically reduce the cost of material irradiation studies and reduce the radiation exposure of personnel conducting the post irradiation examination work.

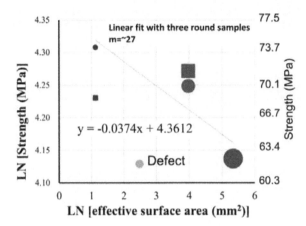

Figure 6. Flexural strength versus effect surface area for sample with different sizes and shapes

CONCLUSION

Four different sizes of round shaped and two different sizes of square shaped graphite samples were used to determine the size effect on the equibiaxial flexural test. The 12 mm diameter round shaped samples showed lower equibiaxial flexural strength and weibull modulus presumably because of specimen machining or fixture defects. The majority of the samples showed average equibiaxial flexural strength of 62-75 MPa and weibull modulus of 30-38. The weibull scaling law is proved to be applicable across different specimen sizes for the round shaped samples. It does not apply to square shaped samples possibly because of alignment issues associated with square specimens. The obtained equibiaxial flexural strength appears to be consistent with the four point bend flexural test. Equibiaxial flexural test with small specimens is proved to be useful and reliable.

ACKNOWLEDGEMENT

This work is sponsored by Ibiden Co. LTD, Japan under contract number NFE-11-03389. Work was carried out at the Oak Ridge National Laboratory for US Department of Energy under Contract DE-AC05-00OR22725 with UT-Battelle, LLC.

REFERENCES
[1]R. E. Nightingale, "Nuclear Graphite." Academic Press: New York, (1962).
[2]S. Ishiyama, T. D. Burchell, J. P. Strizak, and M. Eto, "The Effect of High Fluence Neutron Irradiation on the Properties of a Fine-Grained Isotropic Nuclear Graphite," *J. Nucl. Mater.,* **230**[1] 1-7 (1996).
[3]B. T. Kelly, "Physics of Graphite." Applied Science: London, (1981).
[4]S. Kondo, Y. Katoh, J. W. Kim, and L. L. Snead, "Validation of Equibiaxial Flexural Test for Miniaturized Ceramic Specimens." in DOE/ER-0313/46. 2009.

[5]M. P. Hindley, M. N. Mitchell, D. C. Blaine, and A. A. Groenwold, "Observations in the Statistical Analysis of NBG-18 Nuclear Graphite Strength Tests," *J. Nucl. Mater.*, **420**[1-3] 110-15 (2012).

[6]B. C. Mitchell, J. Smart, S. L. Fok, and B. J. Marsden, "The Mechanical Testing of Nuclear Graphite," *J. Nucl. Mater.*, **322**[2-3] 126-37 (2003).

HIGH TEMPERATURE STEAM CORROSION OF CLADDING FOR NUCLEAR APPLICATIONS: EXPERIMENTAL

Kevin M. McHugh, John E. Garnier, Sergey Rashkeev, Michael V. Glazoff, George W. Griffith, and Shannon M. Bragg-Sitton, Idaho National Laboratory, Idaho Falls, ID, USA

ABSTRACT

Stability of cladding materials under off-normal conditions is an important issue for the safe operation of light water nuclear reactors. Metals, ceramics, and metal/ceramic composites are being investigated as substitutes for traditional zirconium-based cladding. To support down-selection of these advanced materials and designs, a test apparatus was constructed to study the onset and evolution of cladding oxidation, and deformation behavior of cladding materials, under loss-of-coolant accident scenarios. Preliminary oxidation tests were conducted in dry oxygen and in saturated steam/air environments at $1000^{\circ}C$. Tube samples of Zr-702, Zr-702 reinforced with 1 ply of a β-SiC CMC overbraid, and sintered α-SiC were tested. Samples were induction heated by coupling to a molybdenum susceptor inside the tubes. The deformation behavior of He-pressurized tubes of Zr-702 and SiC CMC-reinforced Zr-702, heated to rupture, was also examined.

INTRODUCTION

A primary requirement of advanced nuclear fuel cladding designs is to contain the nuclear fuel inside the fuel rods in the event of a loss-of-coolant accident (LOCA) without diminishing heat transfer characteristics or impairing coolant flow during normal operation. This critical function would increase the safety margin of nuclear reactor designs significantly. While Zr-based alloys are currently used as cladding in most nuclear reactor designs, a variety of advanced cladding materials including metals, ceramics and metal/ceramic composites are being considered as replacements. The corrosion behavior of these materials depends on many factors, including the chemical nature of the cladding material, the reaction temperature, and the type and concentration of the oxidizing agent. For example, corrosion of Zr in air follows a different oxidation pathway than it does in steam. In air, the predominant reaction is:

$$Zr\ (s) + O_2(g) \rightarrow ZrO_2\ (s).$$

In steam, oxidation is accompanied by the evolution of hydrogen gas which can potentially ignite:

$$Zr(s) + 2H_2O(g) \rightarrow ZrO_2\ (s) + 4H_2\ (g)$$

During a LOCA, zirconium fuel cladding can rapidly heat up to temperatures exceeding $1000^{\circ}C$ due to redistribution of fuel thermal energy and fission product decay heating. Steam/cladding interactions above the α/β transformation temperature ($\sim815^{\circ}C$) result in increased oxygen transport inside the metal, the formation of the brittle oxygen-stabilized α phase, and growth of a brittle ZrO_2 film on the outside of the cladding. The metal can swell and rupture or fracture during refill quenching, exposing the reactor's primary coolant loop to the fuel.

A number of techniques have been developed over the years to evaluate the oxidation behavior of Zr-based cladding materials in steam and air/oxygen. Early ductility studies by

Hobson and Rittenhouse[1], and metallographic analysis by Pawel[2], led to the adoption of the 17% oxidation criterion, the 1204°C criterion, and Baker-Just oxidation rate correlation.[3,4] These studies examined oxygen uptake above the $(\alpha + \beta)/\beta$ transformation temperature and the formation of the brittle oxygen stabilized α phase.

Silicon carbide exhibits excellent radiation stability, is composed of low activation elements, and retains its strength and shape under high radiation dose conditions. It is used in advanced nuclear fuel element designs to provide high thermal conductivity containment and structural support to oxide-based fuel. Tristructural Isotropic (TRISO) coated fuel particles used in gas-cooled reactors contain one layer of SiC and three layers of pyrolytic carbon surrounding the fuel kernel. Here the SiC acts as a pressure vessel for fission products, providing structural integrity while retaining solid fission products inside the fuel pellet.[5] All-SiC fuel rod cladding designs are in the development and testing stage for use in light water reactors.[6] Zr-based fuel rods over-braided with SiC ceramic matrix composites could also provide additional structural support, increasing the margin of safety of the metal fuel rods (Figure 1).[7]

SiC is more stable than Zr in air and steam environments, and provides the potential for greater structural stability in a LOCA. Researchers have studied the corrosion behavior of high purity CVD SiC in steam and oxygen and have found that corrosion behavior differs substantially.[8] Corrosion in dry oxygen involves the formation of a relatively stable SiO_2 layer:

$$2SiC(s) + 3O_2(g) \rightarrow 2SiO_2 + 2CO(g)$$

Further oxidation is limited by the oxygen diffusion rate through silica. This layer thus protects the underlying SiC, as the diffusion rate is low. In contrast, the introduction of water vapor allows hydrolysis of SiO_2 to occur, and corrosion of SiC proceeds more rapidly following a two-step oxidation/volatilization reaction[9]:

$$SiC\ (s) + 3H_2O(g) = SiO_2(s) + 3H_2(g) + CO(g) \qquad (1)$$

$$SiO_2\ (s) + 2\ H_2O\ (g) = Si(OH)_4\ (g) \qquad (2)$$

Since the $Si(OH)_4$ is volatile, the recession rate of SiC is greater. The kinetics of the oxidation reaction are described by a parabolic rate constant, while the kinetics of volatilization are described by a linear rate constant. As the partial pressure of water vapor increases, the corrosion rate increases.[10] The SiC recession rate accelerates with rising temperature.

The potential benefits of advanced ceramic and composite materials in the design of nuclear reactor components underscores the need to contrast corrosion behavior of these materials with conventional zirconium-based materials in LOCA scenarios under identical test conditions.

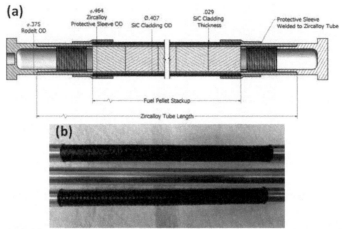

Figure 1. (a) Cladding assembly design, and, (b) fabricated samples of Zircaloy-4 overbraided with 1-ply and 2-ply SiC/SiC ceramic matrix composites (CMCs).[11]

EXPERIMENTAL

The test apparatus used in this study, referred to as the Oxidation Kinetics System (OKS), has been described previously.[12] The system tracks and catalogs outputs from various pressure, temperature and gas composition sensors while exposing heated fuel cladding samples to water/steam, water/air or other oxidation environments. One-color and two-color pyrometers, in conjunction with a thermal image system and thermocouple signals, track the onset of oxidation via emissivity changes at the surface of the cladding. These instruments also sense the exothermic/endothermic reactions that accompany oxidation and phase changes. Gas-phase reaction products of oxidation (e.g., H_2 and CO) are monitored and correlated with temperature/emissivity changes. In addition, nonvolatile or water-soluble oxidation products are collected in a condenser and evaluated off-line.

Te following samples were oxidized in air that was saturated with steam inside a sealed chamber:

1. Closed Zr-702 tubes, pressurized with He to 3.4 MPa and heated at a rate of 37°C/min to bursting.
2. Closed Zr-702 tubes with 1" sleeves of 1ply SiC/SiC CMC. These composite tubes were pressurized with He to 3.4 MPa and heated to bursting at a rate of 37°C/min.
3. Open tubes of Zr-702 heated to 1000°C at a rate of 37°C/min and held for one hour.

Additional samples were tested in dry oxygen. These samples were heated at a rate of about 5°C/min in argon to 1000°C and exposed to dry oxygen for 10 minutes:

1. Open tubes of sintered α-SiC.
2. Open tubes of Zr-702.

The Zr-702 tubes were vertically oriented and fixed at one end. The other end was attached to a pressurized helium gas source via a flexible stainless steel 316L gas line. Helium used to pressurize the samples was ultra high purity. High purity dry oxygen was used in the dry oxygen experiments. High purity (nanopure) water having chemistry characteristics similar to the primary coolant loop of INL's Advanced Test Reactor (ATR) was used to generate steam inside the closed system. The water's pH = 5.1, dissolved oxygen = 5.5 ppm, and conductivity = 3.2 μS/cm were measured using temperature compensated probes for pH (Rosemont Analytical 3500VP sensor), dissolved oxygen (Rosemont Analytical HX438 sensor) and conductivity (Rosemont Analytical 400 sensor).

The CMCs were fabricated by polymer infiltration and pyrolysis (PIP) of a 1-ply woven fabric of Hi Nicalon using polycarbosilane.[13] Both the fibers and the matrix were β-SiC. Commercial pressureless-sintered α-SiC tube samples were formed from ~6 μm power to 98% density using boron and carbon sintering aids.[14] Figure 2 compares the surface appearance of the SiC samples used in this study.

Figure 2. SiC tube samples used in this study. (a) CMC sleeve formed from woven β-SiC fibers that was used to reinforce Zr-702 tubes during deformation testing. (b) α-SiC tube.

Commercial Zr-702 tubes (Zircadyne™ 702, UNS Designation R60702) measuring 9.5 mm OD x 7.6 mm ID x 356 mm were used.[15] The alloy contains up to 4.5 wt.% Hf in solid solution, with the following maximum limits of other elements: Fe+Cr (0.20%), H (0.005%), N (0.025%), C (0.05%), and O (0.16%). While the Hf content of the alloy precludes its use in a nuclear reactor, it was chosen due to cost and availability for these preliminary studies.

Samples were induction heated from the inside-out by coupling to a 7.5 mm OD x 254 mm long molybdenum rod placed inside the tubes. The gap between the Mo and cladding was chosen to be similar to the gap between fuel pellets and cladding in a typical pressurized water reactor (PWR). Temperature was measured using type K thermocouples placed 25 mm inside the Mo susceptor, as well as at the center of the Zr tube and 50 mm above the center of the tube.

RESULTS AND DISCUSSION

Oxidation in steam/air and dry oxygen at 1000°C.
    Figure 3 illustrates the surface oxidation of Zr-702 tubing after one hour exposure to steam/air at 1000°C.

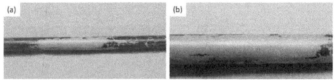

Figure 3. Surface oxidation of Zr-702 tubing following one hour exposure to steam/air at ~1000°C.

The mottled surface was characterized by a white, flaky, fully-converted ZrO_2 product and a tenacious black transitional zirconium oxide layer. Tube samples were sectioned along the length at 25 mm increments from the center of the tube to the edge of the induction heated zone (254 mm). Similar microstructure features were observed for all samples. Figure 4 shows representative SEM photomicrographs of the cross-section of tubing, at various magnifications. Three main microstructure zones were observed as oxygen diffuses into the alloy at 1000°C: an outer, fully-converted zirconium oxide layer; a partially converted underlying zone of oxygen-stabilized α-Zr phase; and a minimally oxidized β-Zr phase of the base metal. The fully converted zirconium oxide layer was about 65 μm thick and was interlaced with axial and transverse cracks of various lengths that resulted from tensile loading due to volumetric expansion. Use of a vCD detector (BSE mode) in Figure 4d clearly revealed these cracks as well as transverse cracks which initiated in the brittle, oxygen-stabilized, α-Zr phase of the metal. The oxygen-stabilized, α-Zr phase was less well resolved but appeared to be about 30 μm thick. Similar trends have been observed by other investigators in steam-oxidized Zircaloy-2 [4, 18] and Zircaloy-4.[16,17, 19] The relative thickness of the fully converted oxide layer and O-stabilized α phase will depend on temperature and time. Weight gain measurements in these studies indicated that the total oxygen uptake for samples tested at 1000°C obeys a parabolic rate law, $W^2 = K_p t$ where K_p is the temperature dependent rate constant, W denotes oxygen weight gain, and t is time. The current study was conducted in acidified steam (simulating a LOCA at INL's ATR reactor). It is not clear to what extent this affected the oxidation behavior.

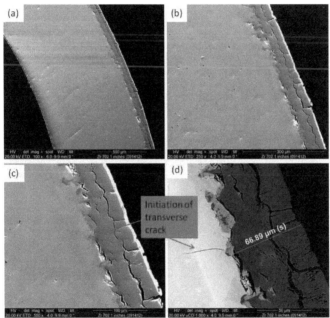

Figure 4. SEM photomicrographs of Zr-702 tubing following one hour exposure in air/supersaturated steam at ~1000°C. (a) 100X, secondary electron mode. (b) 250X, secondary electron mode. (c) 500X, secondary electron mode. (d) 1000X, backscattered electron mode.

EDS element mapping analysis was performed on a cross-section of oxidized tubing to determine the transition in composition from the outer edge of the tube to the interior. Nine scans, each 2.5 min/scan, and evenly spaced by 14.6 μm, were taken progressing from the center of the oxide layer to about 117 μm into the interior of the sample. Figure 5 summarizes concentration data of O, Zr, Cr, Fe, and Hf.

Spot #	O (wt%)	Zr (wt%)	Cr (wt%)	Fe (wt%)	Hf (wt%)
1	0.83	92.3	0.32	0.34	2.13
2	0.79	92.1	0.37	0.43	2.12
3	1.14	92.08	0.32	0.38	2.00
4	1.33	91.78	0.32	0.42	1.79
5	1.87	91.11	0.35	0.40	2.08
6	2.60	90.75	0.30	0.37	1.88
7	3.17	90.16	0.30	0.36	1.86
8	14.37	79.71	0.21	0.27	1.43
9	14.98	79.79	0.03	0.12	1.17

Figure 5. EDS element probe data taken on a Zr-702 sample following one hour exposure to air/supersaturated steam at ~1000°C. (a) SEM photomicrograph showing sampling locations. (b) Wt.% data for O, Zr, Cr, Fe and Hf at various sampling locations.

There is a steep drop in the oxygen concentration from the largely converted ZrO_2 to the O-stabilized α-Zr. This is followed by a gradual decrease in the prior β-Zr region of the interior. The concentrations of Cr, Fe, and Hf in these regions follow a similar pattern. As the layer of O-stabilized α-Zr grows, β-stabilizing elements such as Cr and Fe preferentially diffuse into the β-Zr phase at a high diffusion rate.

Additional samples of Zr-702 and sintered α-SiC were heated to 1000°C in argon and then exposed to high purity, dry oxygen for 10 minutes. Figure 6 compares the microstructures near the edge of a Zr-702 tube (Figure 6a) and pressureless sintered α-SiC (Figure 6b). The Zr-702 sample formed a brittle ZrO_2 layer at the exposed surface measuring about 40 μm thick. Both axial and transverse cracks are present, along with evidence of spalling and delamination. The adjacent oxygen-stabilized α-Zr phase is also about 40 μm thick and appears to be uniform in thickness. Transverse cracks are seen in this layer. Optical microscopy also reveals significant coarsening of the β-Zr phase in the interior of the sample due to the slow heating rate. The α-SiC sample (Figure 6b) has formed a relatively thin layer of SiO_2 that appears to be of variable thickness.

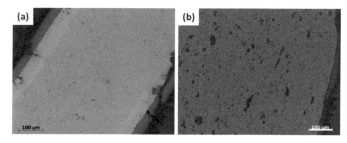

Figure 6. Optical photomicrographs of (a) Zr-702, and, (b) pressureless sintered α-SiC following a 10 minute exposure to dry oxygen at 1000°C.

Figure 7 compares the microstructure of etched Zr-702 samples viewed with polarized light. The starting material, shown in Figure 7a, is characterized by a uniform, relatively fine-grained α-Zr phase, with small, uniformly distributed second- phase particles. During oxidation at 1000°C in steam (Figure 7b) or oxygen (Figure 7c), a relatively coarse O-stabilized α-Zr phase formed. A sharp interface forms with the β-Zr because β phase stabilizers such as Fe and Cr diffuse into the β-Zr, as shown in Figures 7 b and 7c. The slow heating rate for the sample oxidized in dry oxygen caused the β-Zr phase to coarsen.

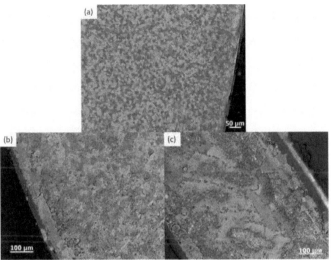

Figure 7. Optical photomicrographs of etched Zr-702 tubing cross sections viewed with polarized light. (a) As-received commercial tubing. (b) Oxidized in steam/air. (c) Oxidized in dry oxygen.

Deformation of pressurized tubes

Nuclear fuel rods used in a P WR are internally pressurized with He to about 2 MPa during manufacture. Helium is added to improve thermal conductivity at the pellet-cladding interface, and to reduce interaction between pellet and cladding during service. As the tubes are heated, the pressure rises. As the fuel ages, gaseous fission products further increase the pressure. The internal pressure is largely offset by the water pressure in the reactor's primary coolant loop. During a LOCA, as primary coolant is lost, the rise in differential pressure can cause the cladding to strain and rupture. Advanced materials such as SiC CMC could increase the reactor's safety margin by providing containment for the nuclear fuel, preventing it from entering the primary coolant.

Deformation of fuel cladding during a LOCA is very complicated. Stress, temperature, and creep strength influence deformation. Azimuthal (circumferential) temperature variations due to fuel movement and eccentricity can play an important role in local temperature and strain behavior. Manufacturing flaws and small differences in microstructure also influence creep. Various studies have been conducted worldwide to study the deformation and ballooning behavior of zirconium alloys in steam, for example, at the PROPAT facility in the UK[20], the EDGAR-1[21] and EDGAR-2[22] facilities in Canada and at CEA in France, the Korean Atomic Energy Institute[23], Japan[24], and elsewhere.

Preliminary experiments were conducted in steam/air on pressurized tubes of Zr-702 measuring 9.5 mm OD x 356 mm long. As-received Zr-702 tubes and Zr-702 tubes reinforced with a ~ 38 mm long tubular sleeve of SiC CMC, as shown in Figure 2a, were tested. The composite tube consisted of single-ply Hi Nicalon β-SiC woven fibers that were polymer-infiltrated with polycarbosilane and pyrolyzed to form a matrix of β-SiC. Tubes were pressurized with He to 3.4 MPa and induction heated to the point of rupture by coupling the induction field to a 254 mm long Mo rod susceptor inserted inside the Zr-702 tubes at the center. Temperature of the susceptor was monitored with a grounded type K thermocouple inserted 25 mm from the end of the susceptor. Tubes were heated at an average rate of $37^{\circ}C/min$.

Figure 8a shows an as-received (starting) Zr-702 tube, and ruptured sample, while Figure 8b shows a starting Zr-702 tube with a 11 mm ID x 11.3 mm OD x 38 mm long SiC CMC sleeve. The sleeve fit loosely over the Zr-702 tube with about 1.5 mm clearance. Since the tubes were vertically oriented, a small ring of a ZrO_2–based adhesive was applied to the Zr-702 tube to center the SiC CMC sleeve and hold it in place. The coating ring appears white in Figure 8b, to the left of the SiC CMC.

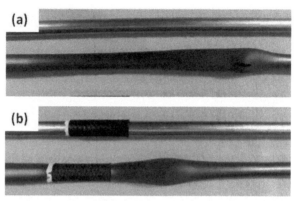

Figure 8. Photographs of starting samples (above) and ruptured samples (below) of (a) Zr-702 tube, and, (b) Zr-702 tube fitted with a SiC CMC sleeve.

Rupture of the unreinforced tube occurred at about $850^{\circ}C$, 13 mm from the center of the tube, and was accompanied by a loud popping sound of tearing metal and rapid pressure drop (Figure 9a). A second unreinforced tube ruptured at $896^{\circ}C$, 93 mm from the center of the tube (Figure 8a). The metal samples ballooned and ruptured in a similar way, with axial tears measuring about 17 mm and 10 mm, respectively. Swelling was observed along the entire length of the tube. Zr-702 in the HCP structure (α phase) transforms to the BCC structure (β phase) at $865^{\circ}C$. The presence of β phase reduces the strength substantially. Similar results have been observed with rupture tests conducted on Zircaloy cladding metal tubes, i.e., rupture occurs as the metal transitions into the $\alpha + \beta$ phase field.

Identical experimental conditions were used to evaluate two samples reinforced with 1-ply SiC CMC sleeves. The samples ruptured at somewhat higher temperatures, about $920^{\circ}C$ and $914^{\circ}C$. The ruptured area of the reinforced samples was small and pin-hole shaped, and was accompanied by a somewhat slower, inaudible release of pressure. The ruptures occurred 5 mm and 7 mm from the center of the tubes, respectively. Notable swelling occurred along the length of the Zr-702 tube except where the tube was reinforced with the SiC CMC sleeve.

Dimensional data for two deformed tubes, one with and one without a reinforcing SiC CMC sleeve, are given in Figures 9 and 10. Diameters were measured in line with the location of the rupture (Figure 9) and at a 90° rotation (Figure 10) for the same samples. Swelling occurred along the entire length of the tubes where they were internally heated. The samples ballooned at the location of the rupture with a maximum circumferential strain of about 80%. This is significant considering that the spacing of fuel rods in a typical PWR design is such that adjacent rods strained by about 32% will touch. At the location of the reinforcing SiC CMC sleeve, metal expansion was significantly constrained. The OD increased about 2%, indicating the 1-ply braided SiC CMC fabric stretched slightly but did not fail under the test conditions.

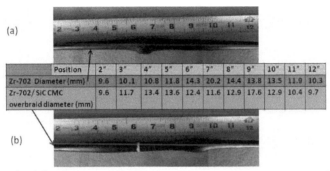

Position	2"	3"	4"	5"	6"	7"	8"	9"	10"	11"	12"
Zr-702 Diameter (mm)	9.6	10.1	10.8	11.8	14.3	20.2	14.4	13.8	13.5	11.9	10.3
Zr-702 / SiC CMC overbraid diameter (mm)	9.6	11.7	13.4	13.6	12.4	11.6	12.9	17.6	12.9	10.4	9.7

Figure 9. Dimensional data for ruptured Zr-702 tubes. Diameter measurements were made in-line with the location of the rupture. (a) Unreinforced Zr-702 tube. (b) Zr-702 tube reinforced with a 38 mm long x 11.3 mm OD SiC CMC sleeve at the center.

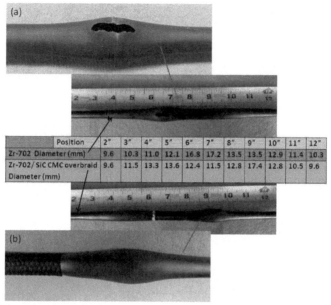

Position	2"	3"	4"	5"	6"	7"	8"	9"	10"	11"	12"
Zr-702 Diameter (mm)	9.6	10.3	11.0	12.1	16.8	17.2	13.5	13.5	12.9	11.4	10.3
Zr-702/ SiC CMC overbraid Diameter (mm)	9.6	11.5	13.3	13.6	12.4	11.5	12.8	17.4	12.8	10.5	9.6

Figure 10. Dimensional data for ruptured Zr-702 tubes. Diameter measurements were made at 90° rotation to the location of the rupture. (a) Unreinforced Zr-702 tube. (b) Zr-702 tube reinforced with a 38 mm long x 11.3 mm OD SiC CMC sleeve at the center. Expanded views at the rupture locations are shown.

CONCLUSIONS

A newly designed test apparatus referred to as the Oxidation Kinetics System (OKS) was used in this study. The system exposes nuclear fuel cladding materials to out-of-reactor, simulated LOCA conditions to study the onset and evolution of oxidation, deformation behavior and quenching behavior.

Preliminary tests were conducted using Zr-702 and SiC ceramic materials. The SiC materials were sintered α-SiC and a β-SiC / β-SiC CMC composed of Hi Nicalon fibers in a matrix of pyrolyzed polycarbosilane. As expected, SiC exhibited improved oxidation resistance over Zr-702 in dry oxygen and steam/air at $1000°C$. Oxygen and steam readily attacked Zr-702 forming a brittle ZrO_2 layer at the exposed surface interlaced with axial and transverse cracks. Oxygen diffusion through this layer resulted in the formation of an oxygen-stabilized α-Zr phase with intermittent transverse cracks. The interior of the Zr-702 transformed to the β-Zr phase which coarsened with exposure time at elevated temperature. The SiC materials formed a thin layer of relatively stable SiO_2.

Preliminary experiments were also conducted in steam/air on pressurized (3.4 MPa He) tubes of Zr-702 and Zr-702 tubes reinforced with a tubular sleeve of SiC CMC. The tubes were induction heated to the point of rupture by coupling the induction field to a Mo susceptor inserted inside the Zr-702 tubes. Rupture occurred as the Zr-702 transitioned into the $\alpha + \beta$ phase field. Ballooning was observed for all samples with a maximum circumferential strain of about 80%. At the location of the reinforcing SiC CMC sleeve, metal expansion was significantly constrained. The OD increased about 2%, indicating the 1-ply braided SiC CMC fabric stretched but did not fail under the test conditions.

ACKNOWLEDGEMENT

The authors gratefully acknowledge contributions of Matt Weseman, Tammy Trowbridge, Todd Morris (INL) and Jesse Johns (Texas A&M University). This work was supported by the U.S. Department of Energy, Office of Nuclear Energy, under DOE Idaho Operations Office Contract DE-AC07-05ID14517.

REFERENCES

[1] D. O. Hobson and P. L. Rittenhouse, "Embrittlement of Zircaloy-Clad Fuel Rods by Steam During LOCA Transients," Oak Ridge National Laboratory, ORNL-4758, January, 1972.

[2] R. E. Pawel, "Oxygen Diffusion in β Zircaloy During Steam Oxidation," *J. Nucl.Mater.* 50 (1974) 247-258.

[3] L. Baker, Jr. and L.C. Just, "Studies of Metal-Water Reactions at High Temperatures. III. Experimental and Theoretical Studies of the Zirconium-Water Reaction," Argonne National Laboratory, ANL 6548, May 1962.

[4] A. F. Brown and T. Healey, "The Kinetics of Total Oxygen Uptake in Steam-Oxidized Zircaloy-2 in the Range 1273-1673 K," *J. Nucl.Mater.* 88 (1980) 1-6.

[5] Lance L. Snead, Takashi Nozawa, Yutai Katoh, Thak-Sang Byun, Sosuke Kondo, and David A. Petti, "Handbook of SiC properties for fuel performance modeling," *J. Nucl. Mater.* 371 (2007) 329-377.

[6] Herbert Feinroth and Bernard R. Hao, "Multi-layered ceramic tube for fuel containment barrier and other applications in nuclear and fossil power plants," Westinghouse Electric, July 2006: WO 2006/076039.

[7] John E. Garnier and George W. Griffith, "Cladding Material, Tube Including Such Cladding Material and Methods of Forming the Same," U. S. Patent Application Attorney Docket 2939-10060US (BA-477).

[8] Nathan Jacobson, Dwight Myers, Elizabeth Opila and Evan Copland, "Interactions of Water Vapor with Oxides at Elevated Temperatures," *J. Phys. Chem. Solids* 66 (2005) 471-478.

[9] Elizabeth J. Opila, "Oxidation and Volatilization of Silica Formers in Water Vapor," *J. Am. Ceram. Soc.*, 86 (2003), 1238-48.

[10] Elizabeth J. Opila, "Variation of the Oxidation Rate of Silicon Carbide with Water-Vapor Pressure," *J. Am. Ceram. Soc.*, 82 (1999), 625-36.

[11] INL/MIS-12-25696 LWRS Fuel Development Plan.

[12] INL/EXT -12-27209.

[13] Physical Sciences, Inc., Andover, MA, USA.

[14] Saint-Gobain Ceramics, Niagara Falls, NY, USA.

[15] Allegheny Technologies, Inc., Pittsburgh, PA, USA.

[16] J. H. Baek, K. B. Park and Y. H. Jeon, "Oxidation kinetics of Zircaloy-4 and Zr-1 Nb-1Sn-0.1Fe at temperatures of 700-1200°C," *J. Nucl.Mater.* 335 (2004) 443-456.

[17] J. H. Baek and Y. H. Jeon, "Steam oxidation kinetics of Zr-1.5 Nb-0.4Sn-0.2Fe-0.1Cr and Zircaloy-4 at 900-1200°C," *J. Nucl.Mater.* 361 (2007) 30-40.

[18] K. Pettersson and Y. Haag, "Deformation and Failure of Zircaloy Cladding in a LOCA. Effects of Preoxidation and Fission Products on Deformation and Fracture Behavior. SKI project B23/78," Studsvik Energiteknik AB, STUDSVIK/K4-80/13, March 1980.

[19] R. R. Biederman, et. al., "A Study of Zircaloy-4 – Steam Oxidation Kinetics," EPRI NP-734 Final Report, April, 1978.

[20] E. D. Hindle and C. A. Mann, "An Experimental Study of the Deformation of Zircaloy PWR Fuel Rod Cladding Under Mainly Convective Cooling," Zirconium in the Nuclear Industry: Fifth Conference, 1980, Boston, MA, USA, ASTM STP 754, ASTM, pp. 284-302.

[21] S. Sagat, H. E. Sills, and J. A. Walsworth, "Deformation and Failure of Zircaloy Fuel Sheaths Under LOCA Conditions," Zirconium in the Nuclear Industry: Sixth International Symposium, 1982, Vancouver, Canada, ASTM STP 824, ASTM pp. 709-733.

[22] T. Forgeron et. al., "Experiment and Modeling of Advanced Fuel Rod Cladding Behavior Under LOCA Conditions: A-B Phase Transformation Kinetics and EDGAR Methodology," Zirconium in the Nuclear Industry: Twelfth International Symposium, 1998, Toronto, Canada, ASTM STP 1354, ASTM pp. 256-278.

[23] J. H. Kim, et. al., "Embrittlement Behavior of Zircaloy-4 Cladding during Oxidation and Water Quench," Nuclear Engineering and Design, 235 (2005) 67-75.

[24] T. Furuta et. al., "Zircaloy Clad Fuel Rod Burst Test Behavior under Simulated Loss of Coolant Conditions in PWRs," *J. Nucl. Sci. Technol.*, 15, (1978), 736-744.

Author Index